Atlas of rock-forming minerals in thin section

W. S. MacKenzie and C. Guilford

Longman

Longman Group Limited
Longman House
Burnt Mill, Harlow, Essex, UK

*Associated companies, branches and representatives
throughout the world*

*Published in the U.S.A. by Halsted Press
a Division of John Wiley & Sons, Inc.*

*First published 1980
Second Impression 1981*

British Library Cataloguing in Publication Data

MacKenzie, William Scott
    Atlas of rock-forming minerals in thin section.
1. Thin sections (Geology)—Pictorial works
I. Title II. Guilford, C
549'.114   QE434

ISBN 0–582–45591–X

Set in 9/10 pt. Monophoto Times New Roman
Printed and bound in Great Britain by
William Clowes (Beccles) Limited, Beccles and London

Contents

Preface iv

Introduction v

Birefringence chart vi

Olivine 1

Monticellite 3

Chondrodite 4

Zircon 6

Sphene 7

Garnet 8

Vesuvanite (Idocrase) 9

Sillimanite 10

Mullite 12

Andalusite 13

Andalusite & Sillimanite intergrowth 15

Kyanite 16

Topaz 17

Staurolite 18

Chloritoid 19

Sapphirine 20

Eudialyte 21

Zoisite 22

Epidote 23

Piemontite 24

Allanite (Orthite) 25

Lawsonite 26

Pumpellyite 27

Melilite 28

Cordierite 30

Tourmaline 32

Axinite 34

Orthopyroxene 35

Augite 36

Titanaugite 37

Clinopyroxene & Orthopyroxene
    intergrowth 38

Aegirine-augite 39

Jadeite 40

Wollastonite 41

Pectolite 42

Anthophyllite – Gedrite 43

Cummingtonite – Grunerite 44

Tremolite – Ferroactinolite 45

Hornblende 46

Kaersutite 48

Glaucophane 49

Arfvedsonite 50

Aenigmatite 51

Astrophyllite 52

Lamprophyllite 53

Muscovite 54

Biotite 55

Stilpnomelane 57

Pyrophyllite 58

Talc 59

Chlorite 60

Serpentine 62

Prehnite 63

Microcline 64

Perthite & Microperthite 65

Sanidine 66

Anorthoclase 67

Plagioclase 68

Quartz 70

Myrmekite 72

Granophyric texture 73

Tridymite 74

Cristobalite 75

Nepheline 76

Sanidine & Nepheline 78

Leucite 79

Nosean 80

Cancrinite 81

Scapolite 82

Analcite 83

Corundum 84

Rutile 85

Perovskite 86

Spinel 87

Brucite 88

Calcite 89

Dolomite 90

Apatite 92

Fluorite 93

Deerite 94

Howieite 95

Zussmanite 96

Yoderite 97

Index 98

Preface

The purpose of this book is to illustrate the appearance of many of the common rock-forming minerals in thin section under the microscope. It is not our intention that it should be used as a substitute for a mineralogy textbook but rather as a laboratory handbook for use in practical classes together with one of the standard textbooks on mineralogy.

The idea of producing a series of photographs of minerals in thin section came from two sources. The son of one of the authors, I. R. MacKenzie, then in his second year as a student of geology, suggested that these would be a useful aid in recognizing minerals under the microscope. On questioning undergraduates in second-year Geology classes in Manchester University, why they preferred certain textbooks to others, the answer was invariably that they found those books which contained illustrations accompanying the text particularly useful, especially when they could recognize under the microscope features which could be seen in the photographs.

Some of the textbooks which, in our opinion, contain the best photomicrographs or drawings of minerals are rather old and are not readily available to the student of today. Rosenbusch's *Mikroskopische Petrographie der Mineralien und Gesteine*, published in 1905, has some excellent photomicrographs printed in black and white, while Teall's *British Petrography*, published in 1888, has beautiful drawings which appear to have been hand-coloured before reproduction by printing. H. G. Smith's "Minerals and the Microscope", first printed in 1914, has been found useful by generations of students of elementary mineralogy because of the high quality of the illustrations. It seemed to us that if we could reproduce faithfully, by colour photography, the appearance of minerals under the microscope both in plane-polarized light and under crossed polars, the usefulness of photomicrographs as a teaching aid would be increased enormously.

The majority of the photographs were made from thin sections of rocks in the teaching collections of the Geology Department in Manchester University and we are grateful to many of our colleagues in Manchester for providing us with thin sections. We are particularly indebted to Professor J. Zussman for his enthusiasm and encouragement to us to undertake this work and to Dr. S. O. Agrell of the Department of Mineralogy and Petrology of Cambridge University who very kindly found, from the Harker Collection in Cambridge, a number of additional thin sections. Dr. Agrell and Professor W. A. Deer very generously agreed to look at most of the photographs we had made and helped us to decide whether they were suitable or could be improved. The authors alone are responsible for any deficiencies which are still present in the photographs. We are also grateful to Dr. J. Wadsworth of the Manchester Department for making a number of useful suggestions for improving the descriptions of the photomicrographs but again we alone are responsible for any errors which may appear here. Finally we are much indebted to Miss Patricia Crook

who typed the text, not once but innumerable times, until we found what we considered to be a compromise between too detailed and too brief descriptions of the photographs.

We should like to thank the staff of the publishers, particularly Miss Bobbi Gouge, for their consideration and helpfulness in the preparation of this work.

Introduction

The minerals represented here are arranged in the same order in which they appear in Deer, Howie and Zussman's *Introduction to Rock Forming Minerals* (relevant page numbers given at the end of each entry in square brackets), except for a few minerals which are not described by these authors, viz. deerite, howieite, zussmanite, yoderite and lamprophyllite. The decision as to which minerals to include has been based mainly on two considerations, firstly, how frequently they occur and secondly whether a photograph can be a useful aid in identification.

In the headings for each mineral we have listed the chemical formula (simplified in some cases), crystal system, optic sign, the values of the β refractive index for biaxial minerals and the ω and ε ray refractive indices for uniaxial minerals together with the birefringence. These figures have been quoted from Deer, Howie and Zussman's book with their permission. The rock type and locality of the specimens are quoted, where these are known, and the magnifications used in taking the photographs are given. Each photograph is accompanied by a brief description of the field of view illustrated but, in general, only properties which can be seen in the photographs are discussed. Thus we have omitted reference to optic axial angle, sign of elongation and dispersion. In most cases at least two photographs have been made for each mineral, one in plane-polarized light and the other the same view under crossed polars. If the mineral is pleochroic we have reproduced two photographs in plane-polarized light with the polarizer in two orthogonal positions. In the case of isotropic minerals we have tended to omit the view taken under crossed polars.

With few exceptions the polarizer has been set parallel to the edges of the photograph but we have not made much use of this fact since discussion of extinction angles is omitted except in the case of the plagioclase feldspars, because this would necessitate reproducing a number of photographs taken under crossed polars. In order to show pleochroism, we have used rotation of the polarizer rather than rotation of the stage of the microscope for two reasons. Firstly, this makes it easier to compare the photographs and observe the change in colour shown by any one crystal and secondly it has been done to encourage the use of this method for detecting weak pleochroism.

Although we have adopted the procedure of retaining the thin section in the same orientation for all three photographs, this has one disadvantage. If there are only a few crystals in the field of view, or the crystals have a strong preferred orientation in the rock section used, we have been unable to show the maximum change in absorption colour on rotation of the polarizer through 90° since the extreme absorption colours are shown by a crystal when its vibration directions are parallel to and perpendicular to the polarizer. In these positions the crystal would be at extinction when viewed under crossed polars and ideally we wish to show the characteristic interference colours near to their maximum intensity. We have not specified in which of the two orthogonal positions the polarizer is set in the photographs taken in plane-polarized light.

As mentioned above we have quoted the numerical value of the birefringence for each mineral, but in the description of the photograph we have generally referred to the order of the interference colour. To enable the reader to translate birefringence to a particular colour we have included on p. vi a photograph of a quartz wedge with a birefringence scale along its length. *This should not be used as a Michel-Lévy chart* since the thickness of the section is not taken into account, it being assumed that the section is of standard thickness, viz. 0·03 mm. Thus the mineral names are reproduced against the highest-order colour which they show in a thin section of standard thickness rather than opposite radial lines which show the variation in colour with thickness and birefringence of the mineral as in a Michel-Lévy chart.

The faithful reproduction of the interference colours of minerals in thin section or in a quartz wedge as seen under crossed polars, depends to a large extent on the type of film used and also on the printing process. Some of the Michel-Lévy charts that have been published depart slightly from the true colours and one fault which is fairly common concerns the middle of the second-order colours where a broad band of bright green is sometimes shown between blue and yellow. Observation of a quartz wedge under crossed polars reveals that the second-order colour between blue and yellow is a rather pale green in contrast to the fairly deep green in the third order. Only in minerals which are colourless and have negligible dispersion, is it possible to distinguish these two greens and even then only after considerable experience. In some of the photographs of minerals of moderate birefringence the edge of the crystals can be seen to be wedge-shaped and thus the order of the interference colour can be determined fairly readily.

Some of the common minerals which are usually considered difficult to identify (e.g. cordierite) are represented by more than one rock section if we considered that additional photographs would give a better idea of the variations in appearance which may be expected in different rocks or if it was impossible, in one field of view, to illustrate the different properties which we wished to show.

In a few cases the photographs taken in plane-polarized light show pale pink and green colours due to stray polarization produced in the photographic equipment: when such colours are present we have noted this in the description of the photograph.

Birefringence chart

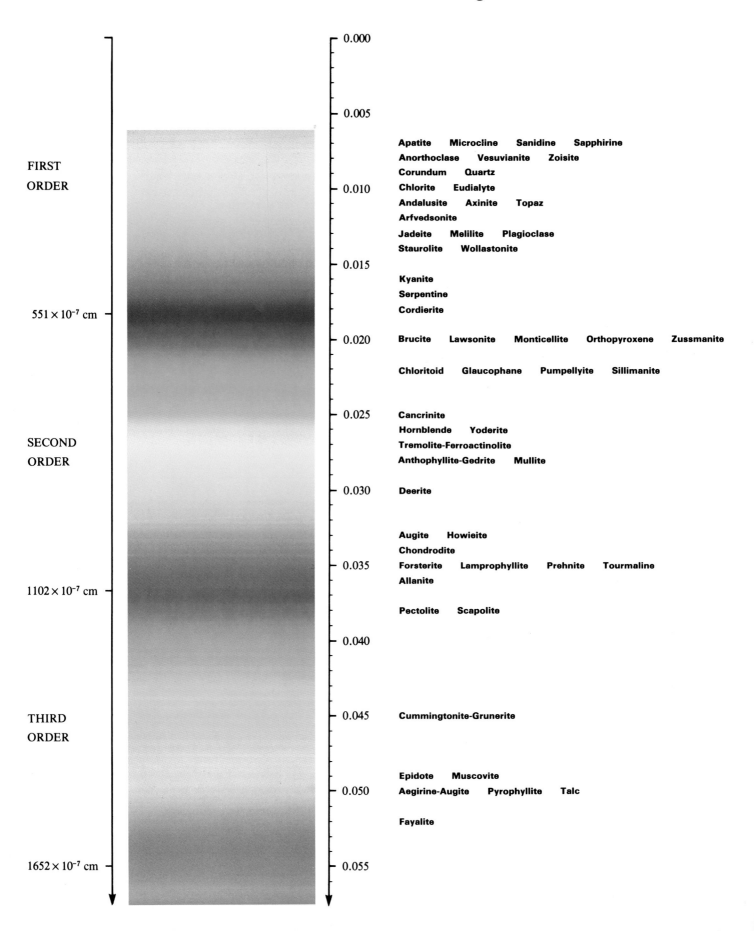

FIRST
ORDER

SECOND
ORDER

THIRD
ORDER

551×10^{-7} cm

1102×10^{-7} cm

1652×10^{-7} cm

0.000

0.005

0.010

0.015

0.020

0.025

0.030

0.035

0.040

0.045

0.050

0.055

Apatite   Microcline   Sanidine   Sapphirine
Anorthoclase   Vesuvianite   Zoisite
Corundum   Quartz
Chlorite   Eudialyte
Andalusite   Axinite   Topaz
Arfvedsonite
Jadeite   Melilite   Plagioclase
Staurolite   Wollastonite

Kyanite
Serpentine
Cordierite

Brucite   Lawsonite   Monticellite   Orthopyroxene   Zussmanite

Chloritoid   Glaucophane   Pumpellyite   Sillimanite

Cancrinite
Hornblende   Yoderite
Tremolite-Ferroactinolite
Anthophyllite-Gedrite   Mullite

Deerite

Augite   Howieite
Chondrodite
Forsterite   Lamprophyllite   Prehnite   Tourmaline
Allanite

Pectolite   Scapolite

Cummingtonite-Grunerite

Epidote   Muscovite
Aegirine-Augite   Pyrophyllite   Talc

Fayalite

Olivine

Mg_2SiO_4–Fe_2SiO_4

Symmetry	=	Orthorhombic $(+)(-)$
RI β	=	1·651–1·869
Birefringence	=	0·035–0·052

The olivines form a complete solid solution between the magnesian end-member, forsterite, and the iron end-member fayalite.

These photographs show two olivine phenocrysts in a fine-grained groundmass of plagioclase feldspar, pyroxene and iron ore. The upper photograph, taken in plane-polarized light, shows the typical shape of olivine crystals; the irregular cracks and slight alteration along the cracks are characteristic of this mineral: there are signs of cleavage along the length of one of the crystals.

In the lower photograph, taken under crossed polars, one of the crystals is cut very nearly perpendicular to an optic axis and so shows a very low interference colour; it is an anomalous brown caused by dispersion of the optic axes. The other crystal shows a second-order blue on the rim whereas the main part of the crystal shows a slightly lower colour. The higher birefringence on the rim of the crystal is an indication of a higher iron content. The reverse effect, viz. lowering of the birefringence colour due to the wedge shape of the crystal boundary, can be seen on the bottom edge of one of the olivine crystals and also on a clinopyroxene phenocryst part of which just appears at the bottom of the field. [1].

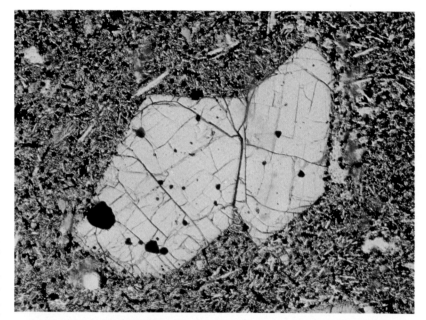

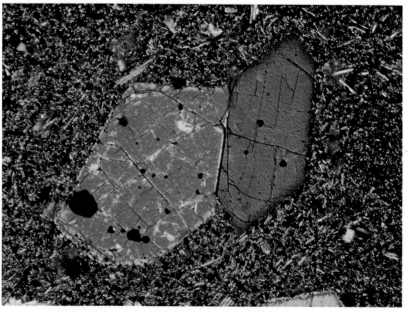

Specimen from ankaramite, Mauna Kea, Hawaii, magnification × 43.

Olivine

Mg_2SiO_4–Fe_2SiO_4

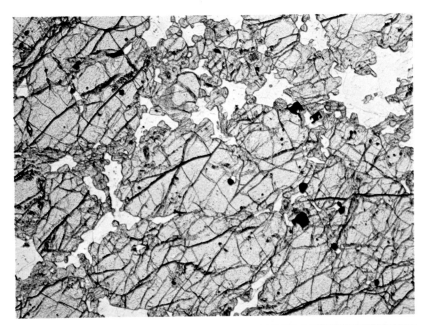

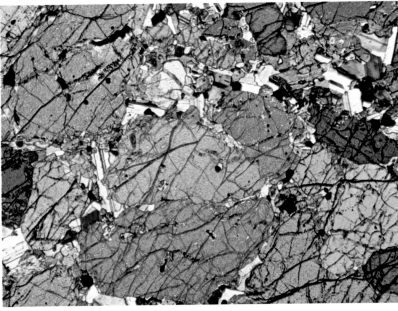

Symmetry		=	Orthorhombic (+) (−)
RI	β	=	1·651–1·869
Birefringence		=	0·035–0·052

The olivines form a complete solid solution between the magnesian end-member, forsterite, and the iron end-member fayalite.

The upper photograph, taken in plane-polarized light, shows olivine (brownish-green colour, occupying most of the field) intergrown with a calcic plagioclase. The high relief of the olivine against the feldspar is noticeable. A pale colour in olivine seen in plane-polarized light is common but it does not show pleochroism – the more iron-rich members of the series show a yellowish-brown colour. The cracks in the crystals are quite characteristic as is the slight alteration of the mineral along the cracks.

In the lower photograph, taken under crossed polars, the interference colours are mostly second order; the highest colour showing in this view is the yellow in the small crystal just above the centre of the field – these colours indicate a magnesium-rich olivine since birefringence colours well into the third order are only seen in olivines with high iron contents.   [1].

Specimen from gabbro picrite, Border Group, Skaergaard intrusion, East Greenland; magnification × 23.

Monticellite

CaMgSiO₄

Symmetry		=	Orthorhombic (−)
RI	β	=	1·646–1·664
Birefringence		=	0·012–0·020

In the upper photograph, taken in plane-polarized light, the dominant mineral is monticellite with subordinate calcite. In plane-polarized light the calcite can be recognized by its good cleavage and twin lamellae. The high relief of the monticellite against the mounting material can be seen at a small hole near the top edge of the slide.

In the lower photograph, taken under crossed polars, the interference colours are seen to be low first order; the highest colour seen here is the orange-yellow colour. It should be remembered that in rocks without quartz or feldspar present it is sometimes difficult to judge the correct thickness of a section and this section may be slightly thin. [10].

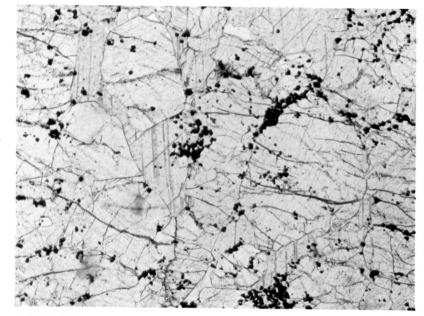

Specimen from monticellite–spinel–phlogopite rock, Barnavave, Carlingford, Eire; magnification × 32.

Chondrodite

Mg(OH, F)$_2 \cdot$2Mg$_2$SiO$_4$

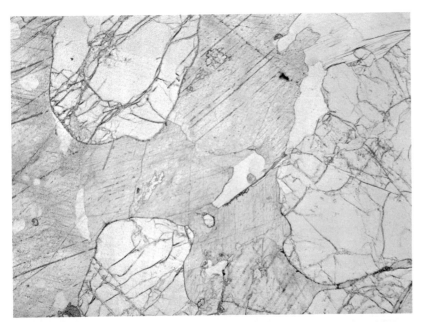

Symmetry	=	Monoclinic (+)
RI β	=	1·602–1·627
Birefringence	=	0·028–0·034

Although members of the humite group, of which chondrodite is one, are frequently yellowish in colour, in this case the chondrodite is nearly colourless in thin section. In the upper photograph, taken in plane-polarized light, the high relief is distinctive, as also is the lack of a good cleavage. Here it is shown intergrown with calcite (brownish colour) and two crystals of muscovite.

In the lower photograph, taken under crossed polars, the muscovite crystals show a bluish-yellow interference colour, while the calcite is grey or dark grey. Multiple twinning is shown in two of the crystals of chondrodite and this is a characteristic of the monoclinic members of the humite group.

It may be that this section is slightly thin because the highest interference colour in this field is the first-order red shown in the crystal in the top right part of the field, and from the birefringence we should expect to see colours up to second-order red (see photographs on p.5). [11].

Specimen from marble, New Jersey, USA; magnification × 20.

Chondrodite

$Mg(OH, F)_2 \cdot 2Mg_2SiO_4$

Symmetry	=	Monoclinic (+)
RI β	=	1·602–1·627
Birefringence	=	0·028–0·034

In this section a concentration of chondrodite, pale yellowish colour under plane-polarized light (upper photograph), is shown intergrown with a garnet (brown). The yellowish colour which characterizes the members of the humite group is very pale here so that a separate photograph to show the pleochroism has not been included. Some of the crystals show signs of poor cleavage.

In the view under crossed polars (lower photograph), the twinning which characterizes the monoclinic member of this series is well illustrated and the interference colours extend up to middle second-order. The garnet in this rock is a grossular and it can be seen to be slightly birefringent.

In addition to chondrodite this rock also contains clinohumite but, since its birefringence is in the same range as that of chondrodite, they can only be distinguished by the fact that the clinohumite has a higher refractive index.   [13].

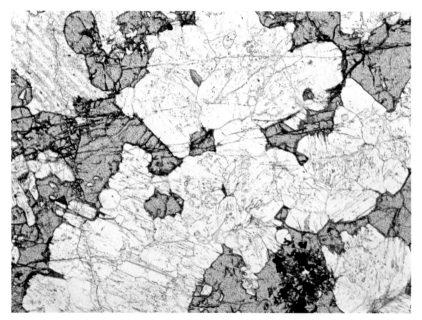

Specimen from marble, Kilchrist, Skye, Scotland; magnification × 28.

Zircon

ZrSiO$_4$

Symmetry		=	Tetragonal (+)
RI	ω	=	1·923–1·960
	ε	=	1·968–2·015
Birefringence		=	0·042–0·065

Zircon commonly occurs in rather small crystals but is easily noticed because of its very high relief. The upper photograph, taken in plane-polarized light, shows rather large zircon crystals associated with sphene in a fine-grained groundmass mainly of feldspar. Sphene also has a very high relief and in this photograph it is very difficult to distinguish from the zircon. The good cleavages in zircon are well displayed in some of the crystals.

In the lower photographs, taken under crossed polars, most of the zircon crystals show high interference colours except for the crystal to the left of centre which shows two cleavages at right angles. This crystal is cut almost at right angles to the optic axis and hence the low interference colours. The sphene crystals can perhaps be more easily identified in this photograph because of their much higher birefringence and by the presence of twinning (top of field of view in centre and to the right of zircon showing low birefringence).   [13].

Specimen from segregation in syenite–pegmatite, Kola peninsula, USSR; magnification × 28.

Sphene

CaTiSiO$_4$(OH, F)

Symmetry	=	Monoclinic (+)
RI β	=	1·870–2·034
Birefringence	=	0·100–0·192

Sphene is a relatively easy mineral to identify because it commonly forms diamond-shaped crystals of very high relief having a brown or red-brown colour. These features are well shown in the upper and middle photographs taken in plane-polarized light. These crystals are pleochroic and simple twinning is common. The mineral intergrown with sphene in this section is alkali feldspar.

The lower photograph, taken under crossed polars, shows a number of crystals near the extinction position but those which show double refraction have a colour very similar to the absorption colour seen under plane-polarized light. This is due to the extreme birefringence of sphene such that the interference colours are very high order, i.e. almost white light. In crystals without the characteristic shape this fact is useful for identification. [17].

Specimen from sphene-rich rock, Kola Peninsula, USSR; magnification × 20.

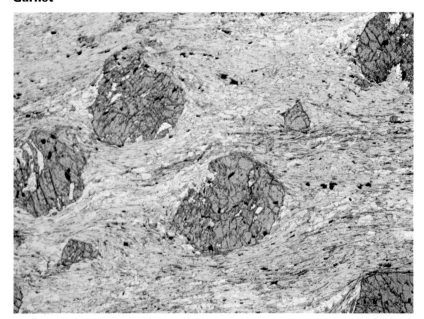

Garnet

$(Mg, Fe, Mn)_3Al_2Si_3O_{12}$
  almandine group
$Ca_3(Al, Fe, Ti, Cr)_2Si_3O_{12}$
  andradite group

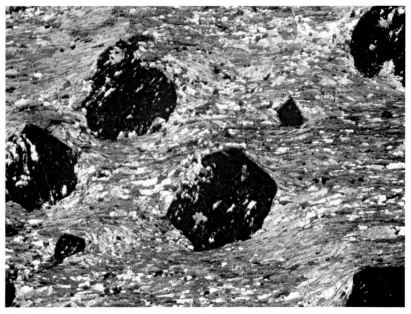

Symmetry	= Cubic
RI n	= 1·714–1·887

A considerable range of compositions is possible in garnets and hence the range of refractive indices quoted. They are very commonly euhedral or subhedral in shape.

The upper photograph shows a number of subhedral garnet crystals, of the almandine series, intergrown with quartz and mica in a metamorphic rock. The garnet stands out quite clearly from the other minerals because of its high relief and brownish colour. It shows inclusions of the groundmass minerals and this is a very common feature.

The middle photograph shows the same view under crossed polars and the garnets are seen to be isotropic (some garnets are birefringent and may show zoning and twinning revealed in the low birefringence colours, see lower photograph, p.5).

The lower photograph, taken in plane-polarized light, shows a melanite garnet (Ti-rich andradite) in an alkaline igneous rock. Its deep brown colour is rather unevenly distributed but it shows zoning at the edges of the crystals: the euhedral shape is very characteristic. The other mineral in this section is altered alkali feldspar.   [21].

Upper and middle specimen from garnet–mica schist, Pitlochry, Scotland; magnification × 11. Bottom specimen from segregation in nepheline syenite, Assynt, Scotland magnification × 20.

Vesuvianite (Idocrase)

$Ca_{10}(Mg, Fe)_2Al_4Si_9O_{34}(OH, F)_4$

Symmetry		=	Tetragonal ($-$)
RI	ε	=	1·700–1·746
	ω	=	1·703–1·752
Birefringence		=	0·001–0·008

In the upper photograph, taken in plane-polarized light, one crystal of idocrase occupies most of the field of view. Its slight yellowish-brown colour can be seen in contrast to a few holes in the section. Its very high relief can also be seen against the mounting medium.

The lower photograph under crossed polars shows the characteristic low anomalous interference colours and the vague signs of bands in the interference colours is also a fairly common feature of large crystals and is an indication of zoning. There is no sign of the poor cleavage in this crystal. The anomalous interference colour is due to strong dispersion and is the most useful property for identifying this mineral: it commonly occurs with grossular garnet which may also show low birefringence colours and sometimes the two minerals are difficult to distinguish.

The green crystals at the lower edge of the field of view and the small green inclusions in the vesuvianite are alkaline amphibole. [32].

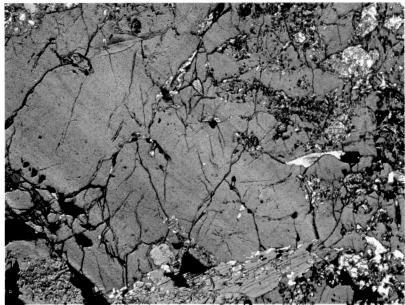

Specimen from unknown locality; magnification × 25.

Sillimanite

Al_2SiO_5

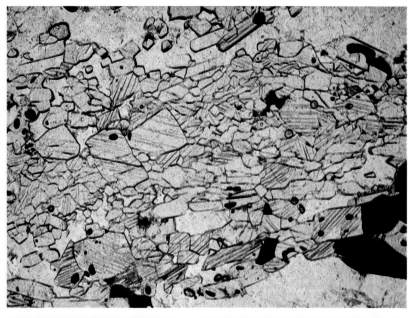

Symmetry	=	Orthorhombic (+)
RI β	=	1·658–1·662
Birefringence	=	0·020–0·022

In the upper photograph, taken in plane-polarized light, all the crystals which stand out in relief are sillimanite which shows clearly against the cordierite with which it is intergrown: at the top left-hand corner of the photograph a yellow halo can be seen in the cordierite (q.v.). In this section the sillimanite has a strong preferred orientation such that most of the crystals are cut at right-angles to their length and show diamond-shaped cross-sections due to the faces of the {110} form; the (010) cleavage is well displayed in some of the crystals.

The interference colours shown in the lower photograph are generally low since the highest colours are shown in crystals cut along the length of the prismatic crystals (see p.11). Crystals cut so that the (010) cleavage is sharp and well defined should be in extinction when parallel to the edges of the photograph. A few crystals cut parallel to their length show second-order colours.   [34].

Specimen from garnet–cordierite–sillimanite gneiss, Ihosy, Madagascar; magnification × 40.

Sillimanite

Al_2SiO_5

Symmetry	=	Orthorhombic (+)
RI β	=	1·658–1·662
Birefringence	=	0·020–0·022

These photographs show lath-like crystals of sillimanite which stand out in high relief against the cordierite with which it is intergrown. Within the cordierite are numerous small needle-like crystals which are also of sillimanite: bunches of very long narrow crystals of sillimanite are termed fibrolite but the concentration of needles is insufficient to justify the use of this term here.

In the lower photograph, taken under crossed polars, the second-order purplish-blue interference colour is near to the maximum colour shown by sillimanite. Sillimanite is difficult to distinguish from mullite (q.v.) but mullite does not often occur in crystals as large as those illustrated here and is restricted to very high temperature contact metamorphic rocks. [34].

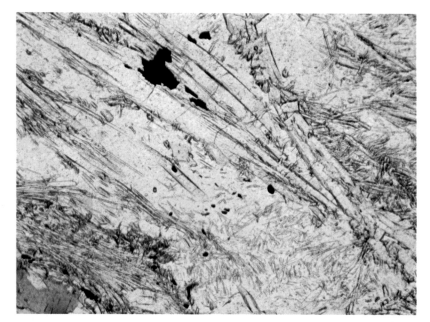

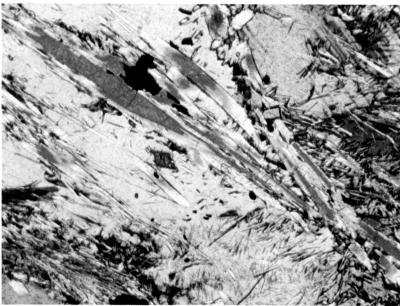

Specimen from cordierite–sillimanite gneiss 11 km south of Ihosy, Madagascar; magnification × 68.

Mullite

$Al_6Si_2O_{13}$

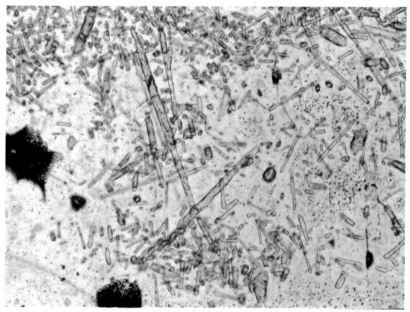

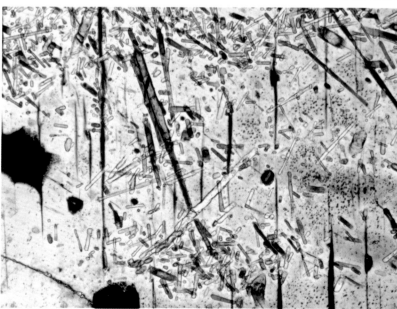

Symmetry		=	Orthorhombic (+)
RI	β	=	1·642–1·675
Birefringence		=	0·012–0·028

Mullite usually occurs in very small needle-like crystals. The upper photograph, taken in plane-polarized light, shows mullite crystals within a large crystal of anorthite (notice the high magnification used for this photograph). These crystals have a very pale pink colour in this section but the pale pink in the background is probably due to stray polarization. The rock is a buchite and the two very dark brown patches in the field of view are of glass.

In the lower photograph, taken under crossed polars, the black lines parallel to the vertical edge of the photograph are plagioclase twin lamellae in the extinction position and in this photograph these could easily be confused with mullite needles. The interference colours shown by these crystals of mullite are not as high as expected and this is caused by the fact that the crystals are thinner than the total thickness of the section. The interference colours of mullite in a section 0·03 mm thick should be about the same as those of sillimanite. [37].

Specimen from buchite, Rudh' a' Chromain sill, Ross of Mull, Scotland; magnification × 164.

Andalusite

Al_2SiO_5

Symmetry	=	Orthorhombic ($-$)
RI β	=	1·633–1·653
Birefringence	=	0·009–0·011

The upper photograph, taken in plane-polarized light, shows one rectangular porphyroblast of andalusite in a fine-grained groundmass. The two cleavages approximately at right-angles to one another can be seen. In the centre of the crystal there is a rectangular area full of inclusions and radiating towards the corners of the crystal there are concentrations of inclusions. This variety of andalusite is known as chiastolite because of the cruciform pattern formed by the inclusions and it is fairly common in low-grade metamorphic rocks.

In the lower photograph, taken under crossed polars, the cruciform pattern is still visible and the characteristic low-order interference colour is seen. [38].

Specimen from chiastolite slate, Lake District, England; magnification × 44.

Andalusite

Al_2SiO_5

Symmetry	=	Orthorhombic $(-)$
RI β	=	1·633–1·653
Birefringence	=	0·009–0·011

In thin section andalusite sometimes shows a pale pink pleochroism and if seen is fairly diagnostic. A pale brownish-pink colour can be seen in the upper photograph, taken in plane-polarized light, and its uneven distribution is characteristic. The high relief against quartz is noticeable. There is a large area in the slide to the right of centre of the photograph where the relief of the andalusite is also obvious – this is a hole in the thin section.

The absorption colour affects the low first-order interference colour so that the same patchy distribution is visible under crossed polars (lower photograph). Andalusite has two good cleavages, (110) and (110), and so most sections show at least one good cleavage. The bright interference colours in this photograph are due to small crystals of muscovite.   [38].

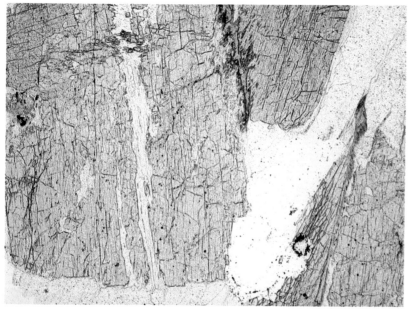

Specimen from contact rock, Ardara pluton, Donegal, Ireland; magnification × 20.

Andalusite & Sillimanite intergrowth

In the upper photograph, taken in plane-polarized light, most of the field is occupied by one crystal of andalusite with a few inclusions of biotite and of quartz. The {110} cleavages of the andalusite can be seen approximately at right-angles to each other and parallel to the edges of the photograph. Intergrown with the andalusite are numerous diamond-shaped crystals of sillimanite with the (010) cleavage bisecting the angle between the andalusite cleavages. The two minerals were probably formed at the same time and the coincidence of the z axes of the crystals reflects the similarity of their structures. The difference in refractive indices of the two minerals is not sufficiently great to show much difference in relief in this photograph.

Under crossed polars (lower photograph) both the andalusite and sillimanite show low interference colours but the colour of the sillimanite is lower than that of andalusite despite the fact that sillimanite has a greater birefringence than andalusite (the vibration directions of both minerals are at 45° to the edges of the photograph). In this orientation both minerals show centred acute bisextrix interferences figures. [34][38].

Specimen from contact rock, Bendoran Cottage, Ross of Mull, Scotland; magnification × 52.

Kyanite

Al_2SiO_5

Symmetry		=	Triclinic $(-)$
RI	β	=	1·721–1·723
Birefringence		=	0·012–0·016

In these photographs kyanite occurs along with quartz, and biotite. In the upper photograph taken in plane-polarized light, kyanite is easily recognized by its high relief compared with quartz and by the fact that the well-developed cleavage appears very dark. It has been suggested that the appearance of kyanite resembles a steel ruler with black engraving marks, since the prominent parting parallel to (001) is frequently visible almost at right-angles to the length of the crystals.

In the lower photograph, taken under crossed polars, the bright interference colours of the biotite contrast with those of kyanite. Unfortunately this section is very slightly thick so that quartz is showing a yellowish tinge hence all the other grains are showing slightly higher interference colours than expected: one half of a twinned kyanite crystal shows a colour close to the sensitive tint red. [41].

Specimen from kyanite gneiss, Glen Urquhart, Inverness-shire, Scotland; magnification × 16.

Topaz

Al$_2$SiO$_4$(OH, F)$_2$

Symmetry	=	Orthorhombic (+)
RI β	=	1·609–1·631
Birefringence	=	0·008–0·011

The upper photograph, taken in plane-polarized light, is of a topaz-quartz rock. The quartz is full of many tiny inclusions whereas the topaz, which stands out in relief against the quartz, is relatively free from inclusions. The perfect (001) cleavage of topaz is visible in one of the crystals.

In the lower photograph, taken under crossed polars, it is difficult to distinguish the topaz from the quartz since their birefringence is almost identical. The narrow white veins on the borders of the topaz crystals are muscovite and this could be an indication of the presence of topaz since topaz is frequently accompanied by muscovite. [45].

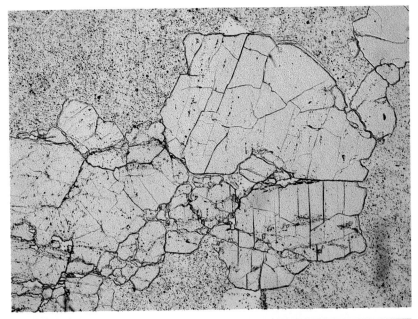

Specimen from topaz–tourmaline–quartz rock, Blackpool Clay Pit, Cornwall, England; magnification × 32.

Staurolite

$(Fe, Mg)_2(Al, Fe)_9Si_4O_{22}(O, OH)_2$

Symmetry	=	Monoclinic (pseudo-orthorhombic) (+)
RI β	=	1·745–1·753
Birefringence	=	0·012–0·014

The upper and middle photographs show porphyroblasts of staurolite with biotite in a fine-grained mass of quartz and feldspar. Pleochroism of the staurolite from yellow to pale yellow is shown in a few of the crystals. The lozenge shape of some of the staurolite crystals is typical and the high relief against the groundmass is well illustrated by the almost black edges of the crystals.

In the lower photograph, taken under crossed polars, the low interference colours can be seen, the large brown-coloured crystal may owe its colour to a combination of the absorption colour and a first-order red. Inclusions as seen here are very common in staurolite. [49].

Specimen from staurolite schist, Waddy Lake, Saskatchewan, Canada; magnification × 20.

Chloritoid

$(Fe, Mg)_2Al_4Si_2O_{10}(OH)_4$

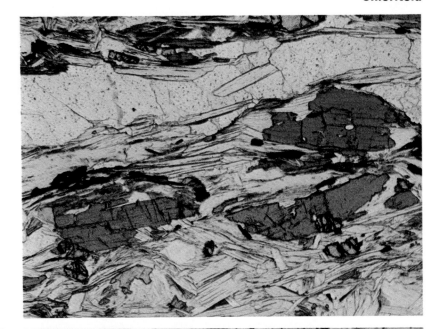

Symmetry	=	Monoclinic or triclinic $(+)$ or $(-)$
RI β	=	1·719–1·734
Birefringence	=	0·006–0·022

The upper and middle photographs, taken in plane-polarized light, show a number of olive-green crystals of chloritoid in which the strong pleochroism to a pale yellow colour can be seen by comparing the two views with the polarizer in orthogonal positions. In this rock chloritoid is intergrown with muscovite and quartz; a few garnet crystals are also visible. The high relief of the chloritoid shows up against the mica but it has lower relief than the garnet. These sections have a preferred orientation and are cut nearly at right-angles to the perfect basal cleavage and hence the strong pleochroism. Sections cut parallel to (001) have only weak pleochroism.

In the lower photograph, taken under crossed polars, the interference colours are slightly anomalous for two reasons, viz. the absorption colours and fairly strong dispersion. No twinning is visible in any of these crystals although chloritoid is commonly multiply-twinned. [52].

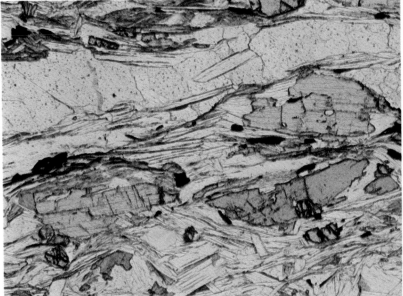

Specimen from schist, Île de Grois, Brittany, France; magnification × 43.

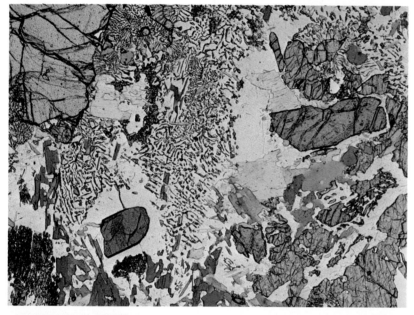

Sapphirine

$(Mg, Fe)_2Al_4SiO_{10}$

Symmetry		= Monoclinic $(-)$ or $(+)$
RI	β	= $1 \cdot 703 - 1 \cdot 728$
Birefringence		= $0 \cdot 005 - 0 \cdot 007$

In the upper and middle photographs the sapphirine crystals are recognized by their colour, which in this case is pleochroic from an indigo-blue to a brownish-yellow colour. In this rock its high relief shows up quite well but there are other high relief minerals in the field, viz. garnet (large crystal at top left-hand corner of the field) and orthopyroxene (pinkish crystals at bottom right). The central part of the field of view is a symplectite intergrowth of cordierite and orthopyroxene. Biotite and quartz are the other minerals present.

Under crossed polars (lower photograph) the interference colours seen in the sapphirine are influenced by the absorption colours and the low birefringence produces anomalous blues.

Careful study of the clear areas reveals yellow pleochroic haloes in the cordierite and the sapphirine crystal in the lower left part of the field of view is surrounded by multiply-twinned cordierite (q.v.).   [57].

Specimen from schist, Val Codera, Italy; magnification ×27

Eudialyte

$(Na, Fe, Ca)_6 ZrSi_6 O_{18}(OH, Cl)$

Symmetry		=	Trigonal $(+)$ or $(-)$
RI	ω	=	1·593–1·643
	ε	=	1·597–1·634
Birefringence		=	0·000 to 0·010

In the upper photograph, taken in plane-polarized light, a number of euhedral crystals of eudialyte show up in relief against analcite with which it is surrounded. In this photograph the substage diaphragm has been left fairly wide open – if it had been more fully closed the relief of eudialyte would show more strongly against the analcite.

The lower photograph, taken under crossed polars, shows two characteristic features of eudialyte which are (*a*) the uneven distribution of low-interference colours, a distribution which is not always clearly related to a growth structure and (*b*) the dark veins of an alteration product. Eudialyte is easily noticed in hand specimen because it is almost always red or brown in colour but in thin section the colour, if present, is generally pale. The crystal at the top left-hand corner of the field is alkali feldspar which here shows a patchy extinction not dissimilar to that of the eudialyte: the three greenish crystals in the field of view are of an alkali amphibole. [59].

Specimen from red kakortokite, Ilimaussaq intrusion, West Greenland; magnification × 42.

Zoisite

$Ca_2Al_3Si_3O_{12}(OH)$

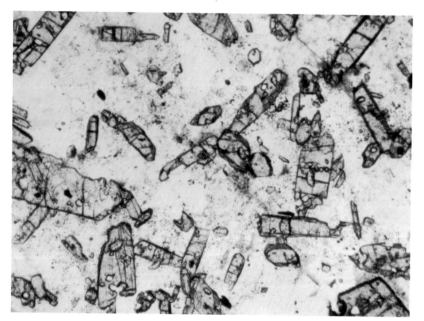

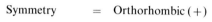

Symmetry		=	Orthorhombic (+)
RI	β	=	1·688–1·710
Birefringence		=	0·004–0·008

The upper photograph, taken in plane-polarized light, shows a number of short prismatic crystals of zoisite intergrown with quartz and a small amount of feldspar. The high relief of the zoisite against the quartz is obvious. There is a suggestion of a cleavage parallel to the length of one crystal at the bottom left.

Under crossed polars (lower photograph) the zoisite crystals show an anomalous blue interference colour somewhat unevenly distributed and this is characteristic of both zoisite and the monoclinic mineral clinozoisite: they are distinguished by the fact that zoisite shows straight extinction in all sections. This sample is zoisite although none of the crystals shown here is exactly parallel to the edges of the photograph.   [61].

Specimen from zoisite schist, Glen Roy, Inverness-shire, Scotland; magnification ×60.

Epidote

$Ca_2Fe^{3+}Al_2Si_3O_{12}(OH)$

Symmetry		=	Monoclinic $(-)$
RI	β	=	1·725–1·784
Birefringence		=	0·015–0·049

The colour of epidote in thin section (yellow or greenish-yellow) is a feature which enables it to be identified fairly readily since the number of common minerals which are yellow in thin section is not great. The pleochroism is shown by comparison of the upper and middle photographs taken in plane-polarized light in which some crystals change from colourless to yellow. The presence of a good cleavage can be seen in a few crystals. The relief of epidote shows against the quartz in the upper right part of the field but the sub-stage diaphragm was rather wide open when these photographs were taken.

The lower photograph shows the same view taken under crossed polars and the bright interference colours, rather unevenly distributed, are very characteristic. [63].

Specimen from epidotized basalt, Michigan, USA; magnification × 32.

Piemontite

$Ca_2(Mn, Fe, Al)_2AlSi_3O_{12}(OH)$

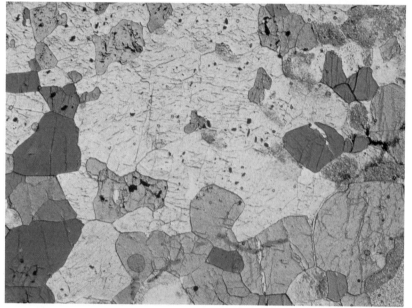

Symmetry		=	Monoclinic (+)
RI	β	=	1·750–1·807
Birefringence		=	0·025–0·088

Although not a very common mineral it is included here because of its spectacular pleochroic colours, viz. yellow, carmine-red and violet or amethyst. This is well shown in the upper and middle photographs taken with the polarizer in orthogonal positions in plane-polarized light: the perfect cleavage does not show up well in this sample nor does the tendency for the crystals to be elongated.

The lower photograph shows that the interference colours are masked by the absorption colours and are thus dominantly red. Two crystals show simple twinning and another shows one thin twin lamella. The intergrown mineral (showing the characteristic amphibole cleavages) is a pale coloured manganese amphibole which although pink in hand specimen is colourless in thin section.

This distinctive pleochroism is not an indication of a particularly high content of manganese: the manganese-bearing variety of epidote called thulite shows the same colours as piemontite. [63].

Specimen from piemontite–quartz rock, Mautia Hill, Tanzania; magnification × 42.

Allanite (Orthite)

$(Ca, Ce)_2 FeAl_2 Si_3 O_{12}(OH)$

Symmetry	=	Monoclinic $(-)$ or $(+)$
RI β	=	1·700–1·815
Birefringence	=	0·013–0·036

The upper photograph, taken in plane-polarized light, shows the brown colour of the mineral which is characteristic as are the dark cracks. The radioactive elements contained in this mineral cause a brown halo in the surrounding rock due to radiation damage but there is no sign of this in the thin section. Allanite is generally slightly pleochroic in shades of brown.

The lower photograph, taken under crossed polars, shows a crystal of biotite enclosed in the larger allanite crystal. These allanite crystals are much larger than are commonly found. The other two minerals visible in this section are microcline (tartan twinning) and quartz. [68].

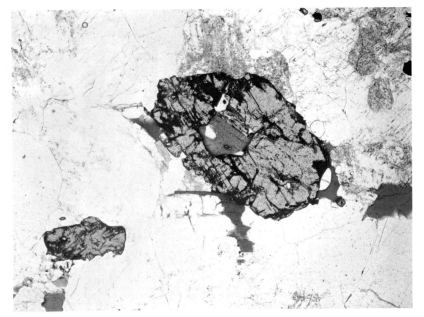

Specimen from granite, near Mandalahy, Madagascar; magnification × 28.

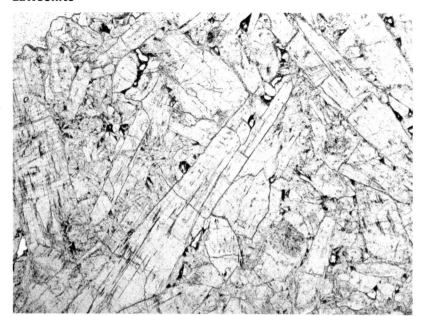

Lawsonite

$CaAl_2Si_2O_7(OH)_2 \cdot H_2O$

Symmetry	=	Orthorhombic (+)
RI β	=	1·674
Birefringence	=	0·020

The upper and middle photographs are of the same field of view, one in plane-polarized light, the other in crossed polars. The field is almost entirely occupied by lawsonite and the feint pleochroism is shown by the slight difference in colour of the crystals in different orientations with respect to the polarizer: many of the crystals show one of the two good cleavages. The birefringence is moderate so the interference colours extend up to second-order and this is shown in the photograph under crossed polars.

The lower photograph made from the same thin section was taken under crossed polars to show multiple twinning in lawsonite (large crystal in the centre of the field). The mineral at the top right corner of this photograph is glaucophane and it is shown here since the occurrence of lawsonite is restricted to glaucophane-schist facies rocks. [70].

Specimen from glaucophane schist, Valley Ford, California, USA. Upper and middle specimen; magnification × 20. Lower specimen with twinning; magnification × 27.

Pumpellyite

$Ca_4(Mg, Fe)Al_5Si_6O_{23}(OH)_3 \cdot 2H_2O$

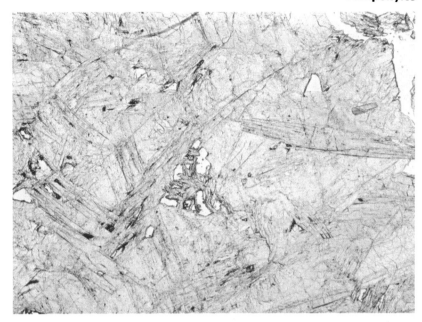

Symmetry	=	Monoclinic (+)
RI β	=	1·675–1·715
Birefringence	=	0·012–0·022

The whole field of view is occupied by pumpellyite except for the clear areas which are voids between crystals and here the high relief of the mineral shows up clearly. The upper and middle photographs, taken in plane-polarized light, show that the mineral is slightly pleochroic from yellow to pale green and this is fairly characteristic of this mineral – the more iron-rich specimens being more deeply coloured. There are signs of cleavages in most of the crystals since pumpellyite has one perfect and one good cleavage.

In the lower photograph, taken under crossed polars, the interference colours range up to second-order blue but the feature which is most noticeable in this photograph is the 'oak leaf' shape formed by the crystals lying diagonally across the centre of the field. This habit is found mainly in vein occurrences of the mineral but when it is seen it is diagnostic. [71].

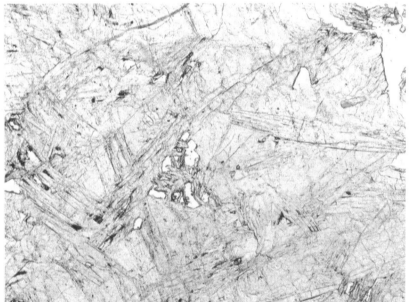

Specimen from glaucophane schist, Tiburon Pass, California, USA; magnification ×44.

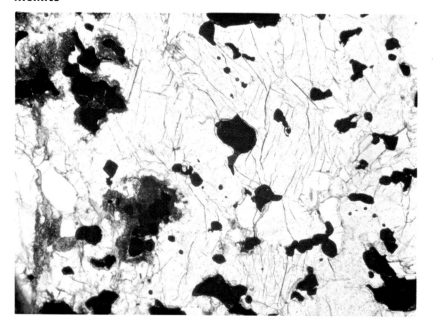

Melilite

$Ca_2Al_2SiO_7$–$Ca_2MgSi_2O_7$

Symmetry	=	Tetragonal $(-)$ or $(+)$
RI ω	=	1·669–1·632
ε	=	1·658–1·640
Birefringence	=	0·000–0·013

The upper and middle photographs are of melilite, in an uncompahgrite, a rather rare melilite-rich rock, one photograph taken in plane-polarized light and the other under crossed polars. In this rock the opaque mineral is magnetite and most of the rest of the field is occupied by melilite. In plane-polarized light there is nothing very distinctive about the mineral and there are signs of at least one cleavage. Under crossed polars however the interference colour is very characteristically an anomalous blue colour and when this can be seen it is a useful diagnostic property taken along with the uniaxial character of the mineral. Zoning of the colour at the edges of the crystals is also fairly typical.

The lower photograph is taken under crossed polars and is from a skarn. Here the zoning from an anomalous blue to an anomalous brown colour is obvious.

The refractive indices quoted above are for the two end-members of the melilite series but the birefringence of zero is for a member of the solid solution containing almost equal amounts of the two end-members. [72].

Upper and middle specimens from uncompahgrite, Uncompahgre, Colorado, USA; magnification × 43. Lower specimen from melilite–phlogopite–clinopyroxene rock, Grange Irish, Carlingford, Ireland; magnification × 72.

Melilite

$Ca_2Al_2SiO_7 - Ca_2MgSi_2O_7$

Symmetry	=	Tetragonal $(-)$ or $(+)$
RI ω	=	1·669–1·632
ε	=	1·658–1·640
Birefringence	=	0·000–0·013

The two photographs shown here are of melilite in an olivine-melilitite, the upper photograph taken under plane-polarized light. In this view the melilite crystals resemble laths of plagioclase in the groundmass of a basalt (with microphenocrysts of olivine) except that the melilite crystals commonly have a dark line along the centre of the laths – seen only in a few crystals in this view – due to included groundmass.

Under crossed polars (lower photograph) the anomalous blue colour zoned to a white colour is quite distinctive and is a fairly certain indication of melilite. The crystals showing yellow, red and a normal blue interference colour are olivine. The brown crystals, best seen in the view in plane-polarized light, at the bottom left of the field are of perovskite. The presence of perovskite is a clue to the occurrence of melilite and vice-versa.   [72].

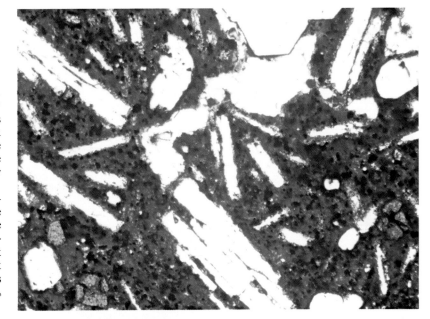

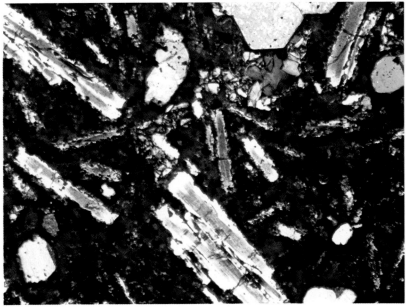

Specimen from olivine–melilitite, Katunga, Uganda; magnification × 83.

Cordierite

$(Mg, Fe)_2Al_4Si_5O_{18}$

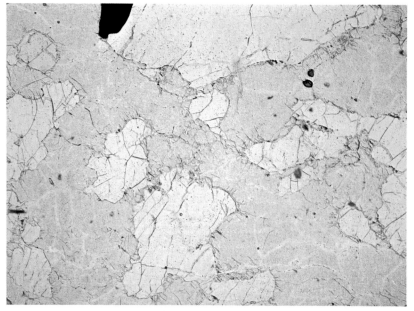

Symmetry		=	Orthorhombic $(-)$ or $(+)$
RI	β	=	1·524–1·574
Birefringence		=	0·005–0·018

The upper photograph, taken in plane-polarized light, shows cordierite intergrown with alkali feldspar. The cordierite can be recognized in this section by its dusty appearance whereas the feldspar is relatively clear. In addition there are irregular cracks and veins at the edges of the cordierite crystals which are yellowish in colour. These are composed of a mineral which is generally called pinite and this alteration is very common in cordierite.

In the lower photograph, taken under crossed polars, the birefringence of the cordierite is seen to be very similar to that of the alkali feldspar but many of the cordierite crystals show lamellar twinning and this causes it to be confused with plagioclase (see photographs on p.31). [84].

Specimen from sillimanite–cordierite gneiss, Fort Dauphin, Madagascar; magnification × 20.

Cordierite

$(Mg, Fe)_2Al_4Si_5O_{18}$

Symmetry	=	Orthorhombic $(-)$ or $(+)$
RI β	=	1·524–1·574
Birefringence	=	0·005–0·018

The upper photograph, taken in plane-polarized light, shows most of the field occupied by cordierite with a number of included minerals showing up in relief (the perfect circles are air bubbles and not mineral inclusions). Around one or two of the inclusions are yellow haloes and these are pleochroic, the mineral forming the halo almost in the centre of the photograph was not in the plane of the thin section.

The lower photograph, taken under crossed polars, shows that most of the field is taken up by a lamellar twinned cordierite. The inclusions which show a second-order blue interference colour are of sillimanite and these do not produce yellow pleochroic haloes.

The presence of yellow pleochroic haloes, frequently seen in cordierite, is one of the most useful diagnostic properties, but these may not be common in cordierites in contaminated igneous rocks in which cyclic twinning is a useful diagnostic feature. [84].

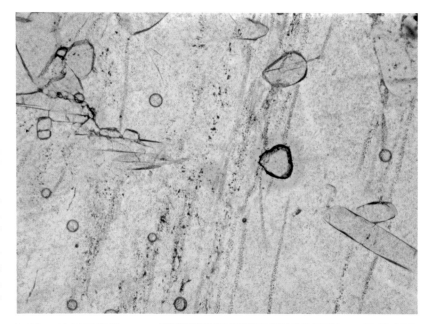

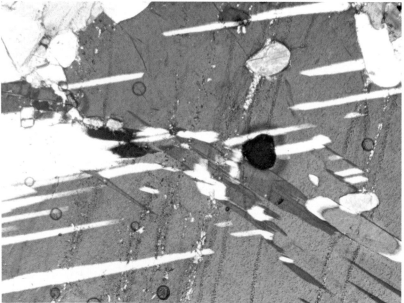

Specimen from cordierite–sillimanite gneiss, near Ihosy, Madagascar; magnification ×72.

Tourmaline

$Na(Mg, Fe)_3Al_6B_3Si_6O_{27}(OH, F)_4$

Symmetry		=	Trigonal $(-)$
RI	ε	=	1·610–1·650
	ω	=	1·635–1·675
Birefringence		=	0·021–0·035

The photographs on this page and the opposite page are of thin sections from the same rock specimen. Comparison of these two photographs, taken in plane-polarized light with the polarizer in orthogonal positions, shows the distinct pleochroism and zoning of the absorption colour. In the lower photograph the polarizer was oriented parallel to the short dimension of the photograph since the maximum absorption colour is shown when the length of the crystal is at right-angles to the plane of polarization of the light. (The crystals were not oriented to show the maximum variation in absorption colour, otherwise in the photograph taken under crossed polars – see next page – they would be in the extinction position.) The other minerals present are quartz and alkali feldspar with crystals of muscovite at the top left and at the right edge of the field. Comparison with the same view, taken under crossed polars, on the opposite page is necessary to identify the muscovite.

Tourmaline shows a wide range of colours in hand specimen; a brownish-yellow, green or blue are the most common colours seen in thin section. In igneous rocks it is usually restricted to late-stage acid varieties but is very common in minor amounts in metamorphosed sediments. The fact that the maximum absorption colour is shown when the length of the crystals is at right-angles to the plane of polarization of the light is a particularly useful diagnostic feature when tourmaline is present in fairly small crystals. [90].

Specimen from topaz–tourmaline–quartz rock, Blackpool Clay Pit, Cornwall, England; magnification × 20.

Tourmaline

Na(Mg, Fe)$_3$Al$_6$B$_3$Si$_6$O$_{27}$(OH, F)$_4$

Symmetry		=	Trigonal $(-)$
RI	ε	=	1·610–1·650
	ω	=	1·635–1·675
Birefringence		=	0·021–0·035

The upper photograph, taken under crossed polars, shows the same field of view as illustrated on the previous page. This shows the moderate birefringence of tourmaline, i.e. up to middle second-order colours.

Since the crystals in this specimen shows a fairly strong preferred orientation another thin section was cut at right-angles to the first and this is shown in the lower photograph taken in plane-polarized light. This view shows the trigonal cross-sectional outline of the tourmaline crystals and again shows zoning of the absorption colours. Sections cut in this orientation can be used to determine the uniaxial negative character of the mineral. [90].

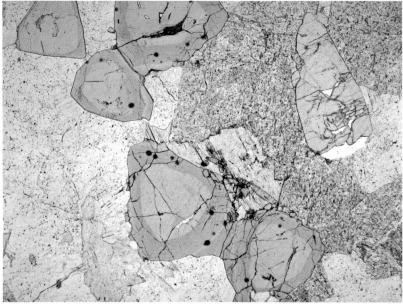

Specimen from topaz–tourmaline–quartz rock, Blackpool Clay Pit, Cornwall, England; magnification × 20.

Axinite

$(Ca, Fe)_3Al_2BSi_4O_{15} \cdot OH$

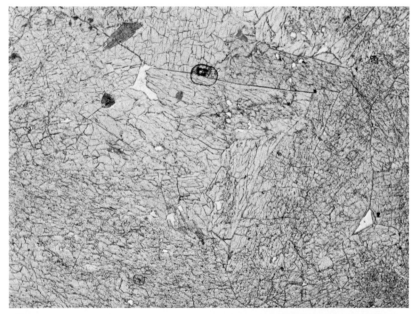

Symmetry		=	Triclinic $(-)$
RI	β	=	1·681–1·701
Birefringence		=	0·009–0·011

The upper photograph, taken in plane-polarized light, shows almost the whole field of view is occupied by axinite which has a pale brownish colour in this specimen but no detectable pleochroism. Most of the crystals show one or more cleavages and this is characteristic since it has four cleavages. The high relief of the mineral shows up against a few holes in the slide.

In the lower photograph under crossed polars, the birefringence is seen to be low. The axe-head shape of the crystals can be more clearly seen under crossed polars and this shape is fairly diagnostic when taken along with the high relief, low birefringence and presence of more than one cleavage in most crystals. [97].

Specimen from axinite–actinolite rock, St. Ives, Cornwall, England; magnification × 32.

Orthopyroxene

$(Mg, Fe)SiO_3$

Symmetry	=	Orthorhombic $(+)$ or $(-)$
RI β	=	1·653–1·770
Birefringence	=	0·007–0·020

In the upper and middle photographs, taken in plane-polarized light, the coloured high-relief mineral is an orthorhombic pyroxene and the most frequently occurring composition is hypersthene. The characteristic pleochroism from green to pink is a good indication of the presence of orthopyroxene. The change in colour may not be very intense but can be detected most easily by rotating the polarizer through 90° as has been done here.

The lower photograph, taken under crossed polars, shows fairly low interference colours. Since the Mg-rich members of this series have the lowest birefringence this is a fairly Mg-rich specimen. The other minerals in this section are quartz, alkali feldspar, plagioclase and one crystal of biotite.

It should be noted that while the pleochroism is a useful diagnostic property it is not always present in orthopyroxenes. [108].

Specimen from charnockite, near Fort Dauphin, Madagascar; magnification × 20.

Augite

$Ca(Mg, Fe)Si_2O_6$

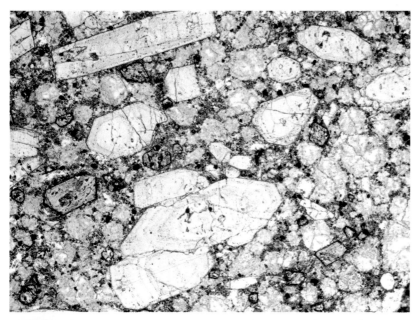

Symmetry	=	Monoclinic $(+)$
RI β	=	1·670–1·741
Birefringence	=	0·018–0·033

The upper photograph, taken in plane-polarized light, shows a number of phenocrysts of augite together with some small feldspars, and a few altered olivines (yellow-brown crystals with black edges) and rounded pseudoleucites. Zoning in the larger phenocrysts of augite is visible in plane-polarized light because of slight differences in absorption colour and density of small inclusions.

Under crossed polars (lower photograph) the zoning is seen more clearly and both simple and lamellar twinning are obvious. The long crystal at the upper left of the field shows hour-glass zoning as well as concentric zoning. The interference colours range up to middle second-order blue.

The black circular regions were probably leucite crystals originally but are now mainly analcite.

The good cleavage usually associated with pyroxenes does not show up in these crystals. This is also true of the pyroxenes in many of the lunar rocks.   [120].

Specimen from pseudoleucite-bearing mafic phonolite, Highwood Mts., Montana, USA; magnification × 20.

Titanaugite

Augites which are rich in titanium usually have a purplish or brown colour in thin section. The upper and middle photographs show crystals of titanaugite taken in plane-polarized light; they have a somewhat darker colour than usual. The pleochroism is quite distinct as also is zoning of the absorption colour. The colourless minerals in this rock are sanidine, nepheline and leucite.

Under crossed polars (lower photograph) the interference colours are, to some extent, masked by the absorption colour but the crystals at the bottom of the field of view show a third-order green colour and this is a higher colour than expected from a normal augite since it represents a birefringence of about 0·04 if the section is of standard thickness. In this view the white area is almost entirely sanidine but the dark region to the top left of the photograph is part of a leucite crystal.   [120].

Specimen from leucite–nepheline–dolerite, Misches, Vogelsberg, Germany; magnification × 20.

Clinopyroxene & Orthopyroxene intergrowth

In the upper photograph, taken in plane-polarized light, the dark crystals are pyroxenes and the light crystals plagioclase feldspars. In some of the pyroxene crystals a lamellar structure can be seen but this is more clearly visible in the lower photograph, taken under crossed polars.

The large crystal just above the centre of the field of view was originally a pigeonite and inverted to an orthopyroxene (dark-brown interference colour) containing lamellae of clinopyroxene (green) parallel to (001) of the original pigeonite: fine lamellae of clinopyroxene (white) are oriented parallel to (100) of the original pigeonite. The crystal at the top left-hand corner of the photograph is a similar inverted pigeonite – the coarse clinopyroxene lamellae are blue but the fine lamellae are only just visible.

The simply-twinned crystal in the lower part of the field of view (violet and reddish-yellow interference colours) is a clinopyroxene twinned on (100) and showing exsolution lamellae of orthopyroxene or pigeonite. [110] [124].

Specimen from norite, Bushveldt intrusion, South Africa; magnification × 24.

Aegirine-augite

$(Na, Ca) (Fe, Mg)Si_2O_6$

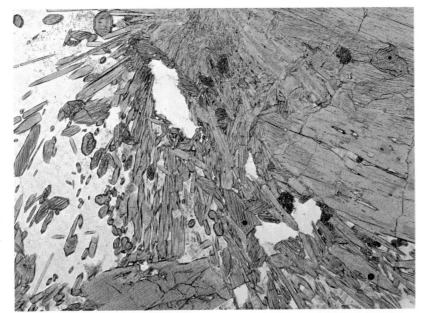

Symmetry   =   Monoclinic $(-)$ or $(+)$

RI   β   =   1·710–1·780

Birefringence   =   0·030–0·050

The upper and middle photographs taken in plane-polarized light show the green to brownish-yellow colour and pleochroism which are diagnostic of sodium-bearing pyroxenes. Zoning shown by variation in the absorption colour is fairly common in such pyroxenes. Most of the crystals show only one of the perfect $\{110\}$ cleavages but a small crystal embedded in the green mass to the right of the field shows both cleavages clearly.

The lower photograph, taken under crossed polars, shows the birefringence associated with this mineral. The distinction between a pyroxene with some of the aegirine molecule ($NaFeSi_2O_6$) and one with a high proportion of the molecule is made on the basis of refractive indices, optic axial angle and extinction angle in an (010) section so that it is not possible from these photographs alone to determine whether this mineral has a small or large amount of the aegirine molecule. [132].

Specimen from sodalite–syenite, Ilimaussaq, West Greenland; magnification × 32.

Jadeite

NaAlSi$_2$O$_6$

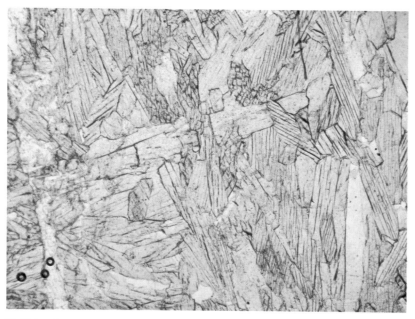

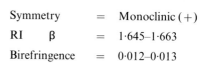

Symmetry		=	Monoclinic (+)
RI	β	=	1·645–1·663
Birefringence		=	0·012–0·013

In the upper photograph, taken in plane-polarized light, two cleavages can be seen in some crystals, the others show one cleavage. There is a slight difference in colour between grains in this section but this is due to stray polarization in the photographic equipment and is not a property of this mineral.

The lower photograph, taken under crossed polars, shows the low birefringence characteristic of jadeite which serves to distinguish it from other clinopyroxenes which have moderate to high birefringence. This section is from a vein in a serpentinite and the whole field is occupied by jadeite.   [137].

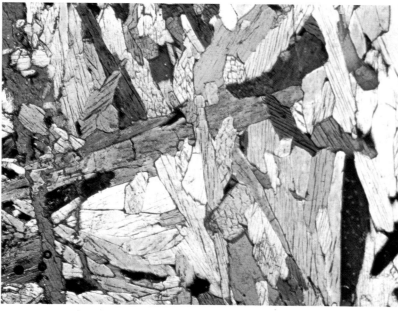

Specimen from jadeite–serpentinite, San Benito Quadrange, California, USA; magnification × 58.

Wollastonite

$CaSiO_3'$

Symmetry	=	Triclinic $(-)$
RI β	=	1·628–1·650
Birefringence	=	0·013–0·014

The upper photograph, taken in plane-polarized light, shows a number of wollastonite crystals lying sub-parallel to the length of the photograph. The other colourless crystals are of nepheline and a few deep green aegirine crystals are also visible. The few slightly cloudy regions (one is almost in the centre of the field) are due to holes in the slide. Wollastonite crystals are elongated along the y crystallographic axes and have three cleavages, all of which are parallel to the y axis so that most crystals show at least one good cleavage.

In the lower photograph, taken under crossed polars, the interference colours extend up to first-order orange but not to red: this is a useful diagnostic property taken along with the tendency for the crystals to be elongated. Simple twinning is common and is shown in the longest crystal.   [140].

Specimen from nepheline–wollastonite rock, Oldoinyo Lengai, Tanzania; magnification × 12.

Pectolite

Ca$_2$NaSi$_3$O$_8$(OH)

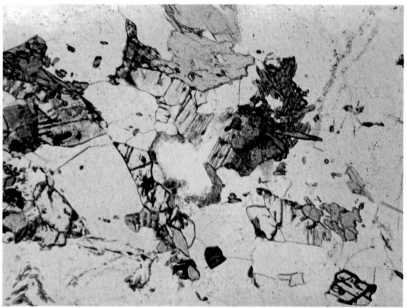

Symmetry		=	Triclinic (+)
RI	β	=	1·605–1·615
Birefringence		=	0·030–0·038

In the upper photograph, taken in plane-polarized light, colourless crystals of pectolite stand out in relief against the other colourless minerals, viz. a sodic plagioclase feldspar and microcline. Pectolite has two perfect cleavages and so the large crystals show at least one cleavage.

In the lower photograph, taken under crossed polars, the pectolite crystals show up clearly because of their bright interference colours extending into second-order, the highest colour in this view being a second-order blue.

The mineral with a slightly green to brown colour in this rock is eckermannite a fairly uncommon amphibole; and a group of small acicular crystals of sodic pyroxene. [144].

Specimen from pectolite–eckermannite–nepheline–syenite, Norra Karr, Sweden; magnification × 53.

Anthophyllite – Gedrite

$(Mg, Fe)_7Si_8O_{22}(OH, F)_2$–

$(Mg, Fe)_5Al_4Si_6O_{22}(OH, F)_2$

Symmetry	=	Orthorhombic $(-)$ or $(+)$
RI β	=	1·605–1·710
Birefringence	=	0·013–0·028

The name anthophyllite is used for the Al-poor members of this series and gedrite for the Al-rich minerals. The upper photograph, in plane-polarized light, shows anthophyllite, biotite and cordierite. The anthophyllite shows considerable relief against the cordierite with which it is intergrown and the typical amphibole cleavages at 120° are seen in some crystals.

The lower photograph, taken under crossed polars, shows that the birefringence is low in comparison with that of most amphiboles. Members of the anthophyllite-gedrite series are orthorhombic in symmetry and in this respect differ from other amphiboles. The absence of twinning is an indication that this may be an orthorhombic amphibole although of course this is not diagnostic. [156].

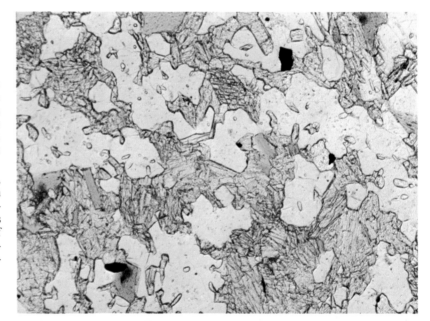

Specimen from cordierite–anthophyllite schist, Pipra, Rewa State, India; magnification × 62.

Cummingtonite – Grunerite

$(Mg, Fe)_7Si_8O_{22}(OH)_2-$
$(Fe, Mg)_7Si_8O_{22}(OH)_2$

Symmetry		=	Monoclinic $(+)$ or $(-)$
RI	β	=	1·644–1·709
Birefringence		=	0·020–0·045

The name grunerite is used for the iron-rich members of this series whereas cummingtonite is used for the intermediate members. No pure Mg end-member is known.

The upper photograph, taken in plane-polarized light, shows cummingtonite crystals intergrown with a plagioclase feldspar and a few crystals of biotite. Unfortunately the characteristic amphibole cleavage does not show up well in this thin section. The dark edges to some of the crystals have a deep olive-green colour which is pleochroic and these are probably due to a common hornblende.

In the lower photograph, taken under crossed polars, interference colours range up to middle second order – the blue-coloured crystals in the centre of the field show the highest colour in this view, i.e. a second-order blue, so that this is a fairly Mg-rich cummingtonite since the colours shown by grunerites extend well into the third order. The fact that the mineral has only a very pale colour in plane-polarized light is also an indication that it is not a grunerite since they tend to be brownish in colour. Multiple twinning is characteristic of members of this series and can be seen in some of the crystals near to the centre of the field of view.   [160].

Specimen from cummingtonite–'norite', Le Pallet, Nantes, France; magnification × 56.

Tremolite – Ferroactinolite

$Ca_2Mg_5Si_8O_{22}(OH, F)_2$
$Ca_2Fe_5Si_8O_{22}(OH, F)_2$

Symmetry	=	Monoclinic (−)
RI β	=	1·612–1·697
Birefringence	=	0·017–0·027

The upper and middle photographs, taken in plane-polarized light, show a group of actinolite crystals showing pronounced pleochroism. Most of the crystals show one good cleavage but none of them are cut to show the angle between the $\{110\}$ cleavages.

The lower photograph, taken under crossed polars, shows twinning in one or two of the crystals and this is fairly common. The interference colours are dominated by greens and browns but this is due to the addition of the absorption colours.

This is actinolite rather than tremolite because of its green colouration since tremolite is colourless.

Since the common twin law for the monoclinic amphiboles is reflection across (100), elongated crystals showing a sharply defined twin junction are the most suitable crystals for measuring the extinction angle $\gamma : z$ since these must be nearly (010) sections. The large twinned crystal showing one half of the crystal in the extinction position is not in a suitable orientation, since the (100) plane is quite oblique to the length of the section and shows up as a white band in this photograph.   [163].

Specimen from unknown locality; magnification × 20.

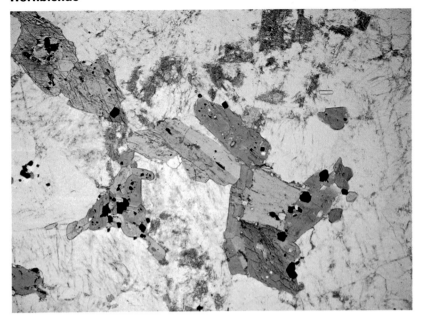

Hornblende

$NaCa_2(Mg, Fe)_4AlSi_6Al_2O_{22}(OH, F)_2$

Symmetry	=	Monoclinic (−) or (+)
RI β	=	1·618–1·714
Birefringence	=	0·014–0·026

The upper and middle photographs, taken in plane-polarized light, show hornblende and biotite together with quartz, alkali feldspar and a sodium-rich plagio-clase. The hornblende shows pleochroism from green to brown whereas the pleochroism of the biotite is from a dark brown to a pale brown. A few of the hornblende crystals show the characteristic shape and two cleavages at 120°.

Under crossed polars (lower photograph) twinning can be seen in a few of the hornblende crystals and the highest interference colour seen in this view is a second-order blue. The difference between biotite and hornblende can be seen in this photograph by the mottled appearance of the interference colours in the large biotite crystal which is near to its extinction position. [167].

Specimen from granite, Moor of Rannoch, Scotland; magnification × 20.

Hornblende

$NaCa_2(Mg, Fe)_4AlSi_6Al_2O_{22}(OH, F)_2$

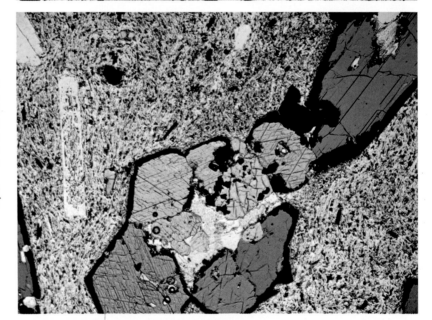

Symmetry	=	Monoclinic $(-)$ or $(+)$
RI β	=	1·618–1·714
Birefringence	=	0·014–0·026

The upper and middle photographs show brownish phenocrysts of hornblende along with plagioclase phenocrysts in a fine-grained groundmass mainly of alkali feldspar. The typical amphibole shape and cleavage can be seen in a few crystals and the pleochroism is quite pronounced. The opaque rims are probably due to the formation of magnetite by oxidation of iron, and are fairly common in hornblendes in volcanic rocks.

The interference colours (lower photograph) tend to be obscured by the absorption colours: the birefringence of common hornblende is low to moderate.   [167].

Specimen from trachyte, Lacqueille, Mt. Dore region, France; magnification × 32.

Kaersutite

$NaCa_2(Mg, Fe)_4(Ti, Fe)Al_2Si_6O_{22}(OH, F)_2$

Symmetry	=	Monoclinic ($-$)
RI β	=	1·690–1·741
Birefringence	=	0·019–0·083

The upper and middle photographs, taken in plane-polarized light, show the strong pleochroism and fox-brown colour which is characteristic of this mineral. The amphibole shape and cleavages are well displayed in the crystal in the centre of the field.

The lower photograph taken under crossed polars, shows that the absorption colours are strong enough to mask partly the interference colours, but the birefringence of this mineral has a very large range and is not a useful diagnostic property. It is difficult to distinguish kaersutite from other brown amphiboles. The amphibole in this rock was previously known as barkevikite. [176].

Specimen from lugarite, Lugar Sill, Ayrshire, Scotland; magnification × 20.

Glaucophane

$Na_2Mg_3Al_2Si_8O_{22}(OH)_2$

Symmetry	=	Monoclinic $(-)$
RI β	=	1·622–1·667
Birefringence	=	0·008–0·022

The upper and middle photographs, taken in plane-polarized light, show mainly glaucophane crystals with a few small quartz crystals. The two prismatic cleavages at 120° are clearly seen and the striking absorption colours which vary from blue to a lavender-blue are typical for this mineral. The colours are zoned near the margins of some crystals.

In the lower photograph, taken under crossed polars, the interference colours are low order, but anomalous because of the strong absorption colours. The zoning is even more easily seen under crossed polars.

The only minerals which show comparable absorption colours are eckermannite which may show a pale lavender colour and yoderite (q.v.) but both are very rare in occurrence. This sample should possibly be described more correctly as crossite since there is generally some Fe^{+3} substituting for Al and the name glaucophane is restricted to minerals with rather low contents of Fe^{+3}. [179].

Specimen from schist, Syphnos, Greece; magnification × 20.

Arfvedsonite

$Na_3(Mg, Fe)_4AlSi_8O_{22}(OH, F)_2$

Symmetry		=	Monoclinic $(-)$
RI	β	=	1·679–1·709
Birefringence		=	0·005–0·012

In the upper and middle photographs, taken in plane-polarized light, arfvedsonite is recognized by its absorption colours which vary from a deep Prussian blue to brownish-green colour. The blue may be so dark that the crystals appear opaque. None of these crystals show the typical amphibole cleavage but the colour and pleochroism are characteristic. In this rock the arfvedsonite is intergrown with alkali feldspar, plagioclase and quartz.

In the lower photograph, taken under crossed polars, the arfvedsonite shows anomalous interference colours, because of the strong absorption colours. The birefringence is low so that bright colours are not to be expected.

It is difficult to distinguish arfvedsonite from riebeckite since both may show the deep blue absorption colour illustrated here, but riebeckite does not show the brownish colour seen in this sample. [187].

Specimen from syenite, Ilimaussaq intrusion, West Greenland; magnification × 32.

Aenigmatite

$Na_2Fe_5TiSi_6O_{20}$

Symmetry	=	Triclinic (+)
RI β	=	1·82
Birefringence	=	0·07

In the upper and middle photographs, taken in plane-polarized light, the very dark brown crystals are aenigmatite. It is sometimes so dark as to appear opaque but the brown colour can usually be seen at the edges of the crystals. A few of the crystals show cleavages and one crystal at the bottom edge of the field has two cleavages at approximately 120°; in this respect it could be mistaken for an amphibole. That it is pleochroic can be seen by comparing these two photographs.

Under crossed polars (lower photograph) the brown colour masks any interference colours so that it is not possible to estimate the birefringence. The large rectangular crystals in this view are sodium-rich alkali feldspars, one of them being in the extinction position. The remainder of the field is occupied by a brownish glass.   [191].

Specimen from pantellerite, Pantelleria, Italy; magnification × 32.

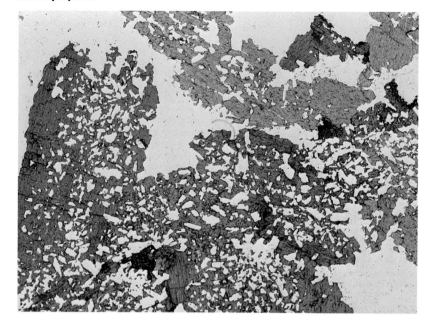

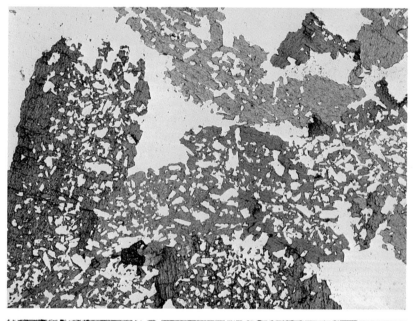

Astrophyllite

$(K, Na)_3Fe_7Ti_2Si_8O_{24}(O, OH, F)_7$

Symmetry	=	Triclinic (+)
RI β	=	1·703–1·726
Birefringence	=	0·06 (approx.)

Astrophyllite usually occurs in needle-shaped crystals in radiating groups, but in this section it occurs as rather shapeless crystals full of inclusions in a fine-grained groundmass. The upper and middle photographs show the pleochroism which is from brown to a yellow colour. Some crystals are much more distinctly yellow than those illustrated. Most crystals show one cleavage.

Because of the strong absorption colours it is difficult to judge the order of the interference colours shown in the lower photograph, taken under crossed polars, but the large crystal at the top of the field shows a third-order green colour which indicates a birefringence of at least 0·04 so we know that the birefringence is fairly high.

The groundmass of this rock consists of a sodium-rich plagioclase feldspar.   [192].

Specimen from microsyenite, East Greenland; magnification × 20.

Lamprophyllite

$Na_3(Ca, Fe)Ti_3Si_3O_{14}(OH)$

Symmetry	=	Monoclinic (+)
RI β	=	1·747–1·754
Birefringence	=	0·032–0·035

This mineral is included because it occurs in some of the rocks from the Pilansberg complex and the Bearpaw Mountains in Montana and these may be represented in many teaching collections.

Lamprophyllite is the pale-brown mineral which shows slight pleochroism (compare upper and middle photographs). A few euhedral crystals can be seen and a cleavage is visible in a number of crystals. (The greenish-coloured minerals are aegirine-augite and arfvedsonite: the colourless mineral is mainly alkali (feldspar).

Under crossed polars (lower photograph) a few of the lamprophyllite crystals show very slightly anomalous colours and one showing hour-glass zoning has a distinctly anomalous brown colour. This crystal is cut nearly perpendicular to an optic axis and the anomalous colour is due to dispersion of the optic axes. Twinning can be seen in one of the crystals. That the blue interference colour is a second-order blue can be readily seen at the wedge-like termination of one of the crystals.

Specimen from green foyaite, Pilansberg, South Africa; magnification × 44.

Muscovite

$KAl_3Si_3O_{10}(OH, F)_2$

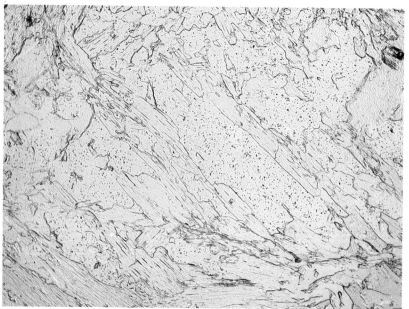

Symmetry		=	Monoclinic (−)
RI	β	=	1·582–1·610
Birefringence		=	0·036–0·049

A very slight greenish tint (upper photograph) can be seen in this muscovite against the quartz with which it is intergrown: this could be due to the mineral being a phengite rather than muscovite but in this case it is due to slight stray polarization in the photomicroscope. Most of the crystals show the perfect basal cleavage.

In the lower photograph, taken under crossed polars, similar interference colours are shown by most crystals due to a preferred orientation in the rock: one crystal near the top right-hand corner of the photograph is cut nearly parallel to the basal cleavage and so gives a good interference figure. Many of the crystals show twinning and crystals which are near to the extinction position show the mottled appearance which is characteristic of all micas. [201].

Specimen from kyanite schist, Hamma of Snarravae, Unst, Shetland, Scotland; magnification × 44.

Biotite

$K(Mg, Fe)_3AlSi_3O_{10}(OH, F)_2$

Symmetry	=	Monoclinic $(-)$
RI β	=	1·605–1·696
Birefringence	=	0·04–0·08

Biotite is invariably brown or green in colour. The upper and middle photographs, taken in plane-polarized light, show a field almost entirely occupied by biotite with numerous dark brown to black pleochroic haloes. The perfect cleavage can be seen in many of the crystals. The maximum absorption colour is shown when the polarizer is parallel to the cleavage.

In the lower photograph, taken under crossed polars, the crystals in the centre of the field are fairly close to the extinction position, and this has been done to emphasize the mottled appearance which is characteristic of all micas and is most clearly seen when close to extinction.

The greenish crystal at the bottom left-hand corner of the field is tourmaline. [211].

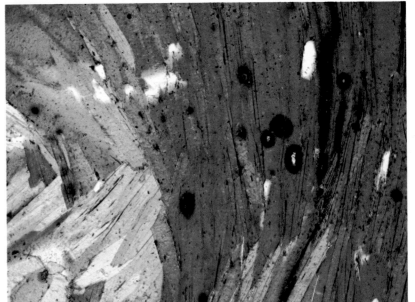

Specimen from biotite–kyanite–gneiss, Dun nan Geard, Ross of Mull, Scotland; magnification × 52.

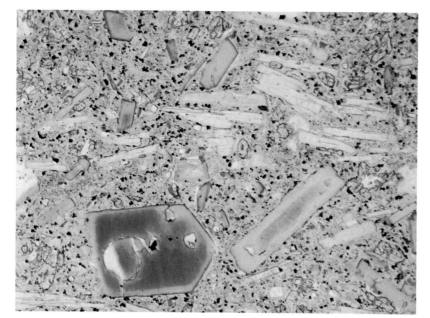

Biotite

$K(Mg, Fe)_3AlSi_3O_{10}(OH, F)_2$

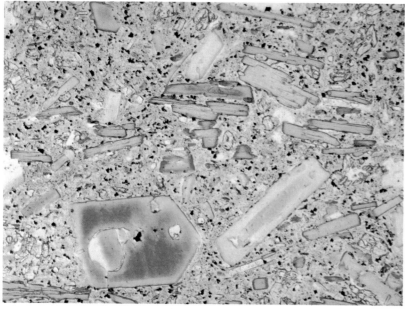

Symmetry	=	Monoclinic ($-$)
RI β	=	1·605–1·696
Birefringence	=	0·04–0·08

These photographs show brown phenocrysts of biotite in a fine-grained groundmass. The change in absorption colours on rotating the polarizer (upper and middle photographs) and zoning of the absorption colours is clearly seen. This zoning is probably due to variation in iron and titanium contents.

In the lower photograph, taken under crossed polars, zoning of the birefringence colours can also be seen. The long crystal to the right of centre shows blues and yellows which are second-order colours.

The other minerals in the rock are feldspars and a few microphenocrysts of pyroxene (high relief).   [211].

Specimen from lamprophyre, Puffin Bay, Herme, Channel Islands; magnification × 32.

Stilpnomelane

$K(Fe, Mg, Al)_3Si_4O_{10}(O, OH)_2 \cdot 3H_2O$

Symmetry	=	Monoclinic $(-)$
RI β	=	1·576–1·745
Birefringence	=	0·030–0·110

The brown lath-shaped crystals in these photographs are of stilpnomelane and the pronounced pleochroism is clearly illustrated by the upper and middle photographs with the polarizer rotated through 90°. It may be brown, as in this example, or green and so can be confused with biotite but the cleavage is not nearly as well developed as in biotites. This sample shows a darker brown colour than is expected in a fresh sample because it is partly oxidized.

The interference colours are generally masked by the brown absorption colour as seen in the lower photograph, taken under crossed polars. There are slight signs of another cleavage at right-angles to the length of the crystals and when this is clear it is a useful observation to distinguish stilpnomelane from biotite.

The chemical formula given above is simplified and does not show the wide variation in composition which this mineral may have. [222].

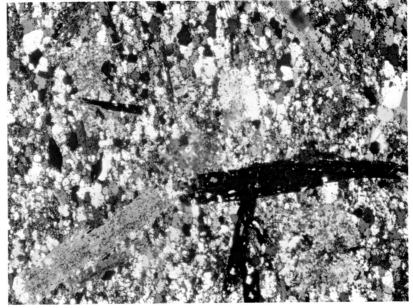

Specimen from metamorphosed ironstone, Laytonville, California, USA; magnification × 32.

Pyrophyllite

AlSi$_2$O$_5$(OH)

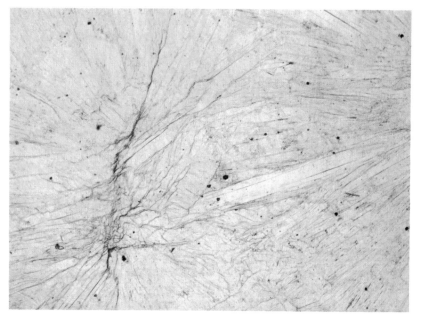

Symmetry		=	Monoclinic ($-$)
RI	β	=	1·586–1·589
Birefringence		=	0·050

The upper photograph shows a field of view almost entirely occupied by pyrophyllite but there are no features visible which permit easy distinction between pyrophyllite and muscovite.

Under crossed polars, lower photograph, the mottled appearance is also similar to that of muscovite. This section was chosen because it shows relatively large crystals of pyrophyllite and an interference figure shows a moderate optic axial angle in contrast to the low value seen in muscovites. Usually pyrophyllite occurs in such small crystals that it is impossible to obtain an interference figure from them. [225].

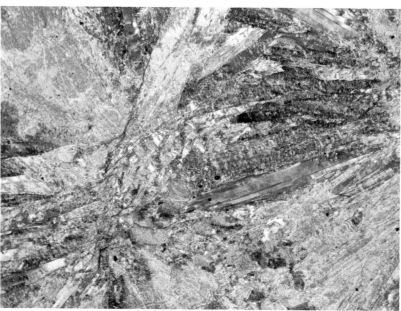

Specimen from unknown locality; magnification × 72.

Talc

$Mg_3Si_4O_{10}(OH)_2$

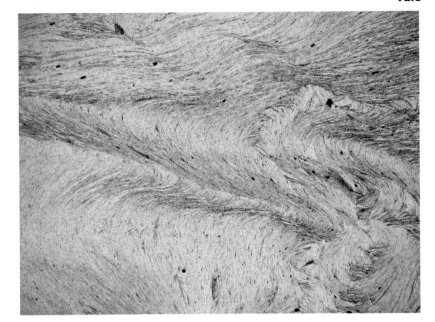

Symmetry	=	Monoclinic $(-)$
RI	β	=	1·589–1·594
Birefringence		=	0·05

Most of the field of view is occupied by talc in a highly deformed rock. It is much easier to identify talc in hand specimen than in thin section because it feels slippery, whereas in thin section it can be confused with a white mica. The change in relief of different parts of the highly contorted bands of talc is well illustrated by the upper and middle photographs, taken with the polarizer in orthogonal positions.

The lower photograph, taken under crossed polars, shows second-order interference colours over almost the whole field except for small areas of chlorite which is intergrown with the talc. The chlorite shows a first-order grey or white colour. [227].

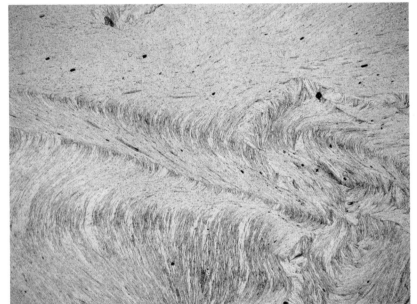

Specimen from Madran Mountain, Menderes Massif, South West Turkey, magnification × 27.

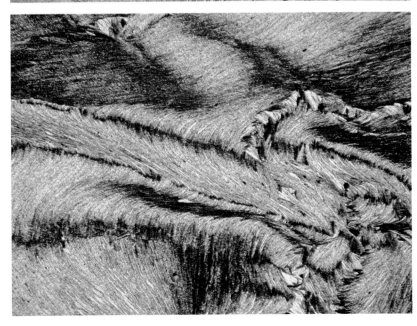

Chlorite

$$(Mg, Fe, Al)_{12}(Si, Al)_8O_{20}(OH)_{16}$$

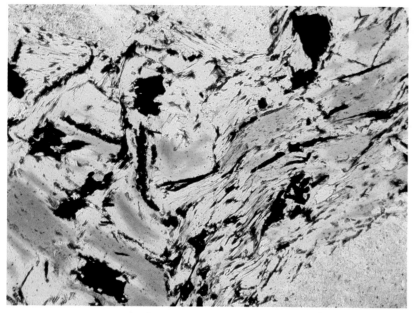

Symmetry		=	Monoclinic (+) or (−)
RI	β	=	1·57–1·67
Birefringence		=	0·00–0·01

The term chlorite covers a wide range of mineral compositions but most members of this group are either colourless or green in colour and when a green mineral is observed intergrown with a brown biotite as in the photograph adjacent it is likely to be chlorite. The upper photograph, taken in plane-polarized light, shows a biotite breaking down to a chlorite so that residual brown flakes of biotite are surrounded by pale green chlorite: it is usually pleochroic. Chlorite has a perfect basal cleavage but it is visible in only a few crystals in this view.

Under crossed polars (lower photograph) the anomalous interference colours which are characteristic of some chlorites are clearly seen. Although all chlorites do not have anomalous colours, their birefringence is always low.   [231].

Specimen from mica–diorite, Glen Loy, Scotland; magnification × 58.

Chlorite

$(Mg, Fe, Al)_{12}(Si, Al)_8O_{20}(OH)_{16}$

Symmetry	=	Monoclinic ($+$) or ($-$)
RI β	=	1·57–1·67
Birefringence	=	0·00–0·01

Only one photograph (upper) taken in plane-polarized light is shown here because there are enough crystals in different orientations to show the pleochroism from pale yellowish to green. The perfect cleavage can be clearly seen in many of the crystals and others are cut nearly parallel to the basal cleavage and, of course, show no cleavage.

In the lower photograph, taken under crossed polars, the low grey and anomalous brown colours are characteristic of some chlorites. There are also signs of twinning in some crystals.

The mineral with which the chlorite is intergrown is adularia, a K-rich feldspar formed in low-temperature veins. [231].

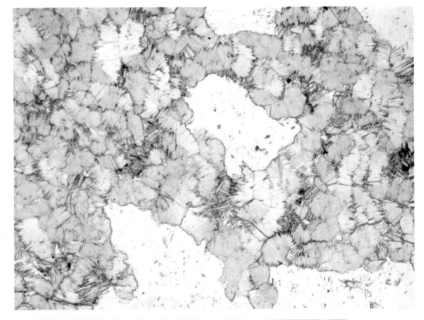

Specimen from adularia–quartz vein, St. Gottard, Switzerland; magnification × 72.

Serpentine

$Mg_3Si_2O_5(OH)_4$

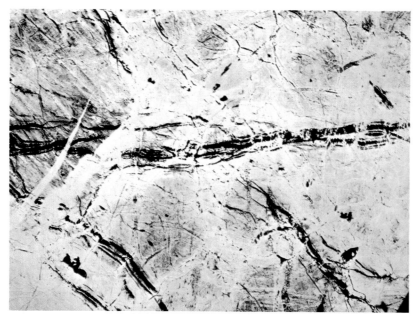

Symmetry		=	Monoclinic ($-$)
RI	β	=	1·54–1·566
Birefringence		=	0·004–0·017

The name serpentine covers three polymorphs which cannot easily be distinguished optically, and many specimens contain more than one polymorphic form. This specimen is probably a mixture of lizardite and chrysotile.

The upper photograph shows serpentine which is pale yellowish in colour together with an opaque iron oxide: this aggregate is undoubtedly the result of breakdown of an olivine or pyroxene, although relict crystal shapes are not clearly defined as in some examples.

The lower photograph, taken under crossed polars, shows the low first-order colours characteristic of the serpentine minerals and it also shows a mesh texture which is a common feature of this mineral and is fairly diagnostic. [242].

Specimen from serpentinite, Lizard, Cornwall, England, magnification × 20.

Prehnite

$Ca_2Al_2Si_3O_{10}(OH)_2$

Symmetry	=	Orthorhombic (+)
RI β	=	1·615–1·642
Birefringence	=	0·022–0·035

Most of the field is occupied by prehnite although a few crystals of calcite can be seen in the top part of the upper photograph, taken in plane-polarized light. The pale pink and green colours are stray polarization colours produced in the photographic equipment. One of the characteristic features of this mineral is its tendency to form radiating groups of crystals and this is best seen in the lower photograph, taken under crossed polars.

Prehnite characteristically shows very bright second- and third-order colours and since the mineral in thin section is almost without colour these interference colours are usually very pure.   [277].

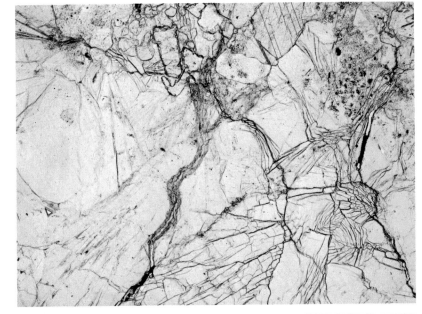

Specimen from marble, unknown locality; magnification × 26.

Microcline

$KAlSi_3O_8$

Symmetry	=	Triclinic ($-$)
RI β	=	1·518
Birefringence	=	0·007

The two photographs show a perthitic microcline crystal cut approximately parallel to (001) with the trace of (010) parallel to the long dimension of the photograph.

The upper photograph was taken in plane-polarized light with the substage diaphragm closed to accentuate the relief in different parts of the section. Parallel to the short edge of the photograph are small veinlets differing in relief from the host – these are microperthitic albite lamellae. At an angle of about 25° to the same edge of the photograph there are three or four thick veins of perthitic albite which also differ in relief from the surrounding material.

The lower photograph, taken under crossed polars, shows the cross-hatched twinning (albite and pericline laws) which is very characteristic of microcline. The perthitic albite veins at 25° to the short edge of the photograph are rather dark here but show twinning according to the albite law – the composition plane (010) is parallel to that of the albite twinned lamellae in the microcline.

The twin lamellae in the microcline are most sharply defined close to the albite veins and they are of variable width. This mineral should not be confused with anorthoclase (q.v.).   [285].

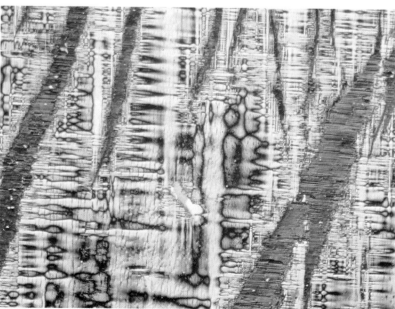

Specimen from pegmatite, Diamond Mine, Topsham, Maine, USA, magnification × 43.

Perthite & Microperthite

(K, Na)AlSi₃O₈

Perthite is the name given to an intergrowth of a potassium-rich and a sodium-rich feldspar when the host material is the potassium-rich feldspar. When the host material is a plagioclase the name antiperthite is used and when the sodium-rich and potassium-rich phases are in equal amounts the term mesoperthite is used. Perthite is used when the intergrowth can be seen in hand specimen and microperthite when it is visible only under the microscope. These three photographs were taken under crossed polars.

The upper photograph shows most of the field occupied by a mineral with a very dark grey interference colour and small white blebs of microperthitic albite fairly uniformly distributed throughout. Two cleavages are visible almost at right-angles to one another so that this section is cut nearly perpendicular to the x axis. No twinning is visible so it is likely to have monoclinic symmetry and the name orthoclase-microperthite is appropriate. A nearly centred acute bisectrix interference figure is obtained from this section with an optic axial angle of about 45°, a value appropriate for orthoclase-microperthite.

The middle photograph is a coarse perthite cut nearly parallel to (010). The white areas are sodium-feldspar and the dark areas are of potassium-feldspar. Within the dark areas are fine light-coloured lamellae of microperthitic albite which lie at an angle of approximately 75° to the trace of the (001) cleavage which is parallel to the long edge of the photograph. Although no twinning can be seen in either the sodium-rich or potassium-rich phases, if albite twinning were present it would not be seen in a section cut nearly parallel to (010). In addition pericline twinning will not be seen in a section cut exactly at right-angles to the twin axis, the y crystallographic axis, and will be difficult to detect in a section close to this orientation. From this section alone it is not possible to say whether the potassium-rich phase is orthoclase or microcline.

The lower photograph is also of a microperthitic feldspar in the same orientation as the middle photograph, but here only microperthitic albite is visible, oriented at about 75° to the trace of the (001) cleavage. [283].

Upper specimen from garnet–granulite, West of Amboasary, Madagascar; magnification × 21. Middle specimen from unknown locality; magnification × 27. Lower specimen from pegmatite, Kodarma, Bihar, India; magnification × 25.

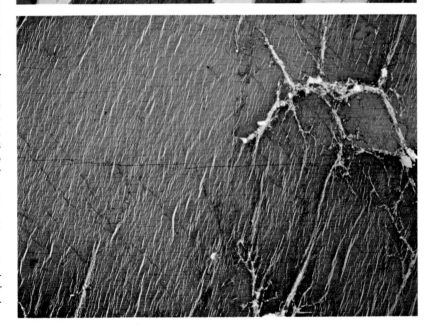

Sanidine

$(K, Na)AlSi_3O_8$

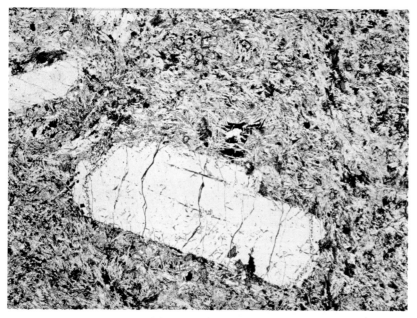

Symmetry		=	Monoclinic $(-)$
RI	β	=	1·522–1·528
Birefringence		=	0·006–0·007

The name sanidine is used for monoclinic alkali feldspars which occur in volcanic rocks; they are usually fairly potassium-rich. The upper photograph taken in plane-polarized light shows a few phenocrysts of sanidine in a groundmass also composed mainly of sanidine. The regular arrangement of inclusions at both ends of the largest crystal outlines the shape of the growing crystal.

Simple twinning, as seen in the lower photograph taken under crossed polars, is very common in monoclinic alkali feldspars and this serves to distinguish them from plagioclases since the latter usually show lamellar twinning as well as simple twinning. The twin law in this case is the Carlsbad law which is the most frequently observed twin law in monoclinic feldspars.   [285].

Specimen from phonolite, San Angelo, Ischia, Italy, magnification × 24.

Anorthoclase

$(Na, K)AlSi_3O_8$

Symmetry	=	Triclinic $(-)$
RI β	=	1·528–1·532
Birefringence	=	0·007–0·008

The name anorthoclase is used for triclinic sodium-rich alkali feldspar which occur in volcanic rocks. The upper photograph taken in plane-polarized light shows a group of crystals of anorthoclase in a fine-grained groundmass of alkali feldspar and quartz. There are slight signs of cleavages because the largest crystal is cut almost at right-angles to both (001) and (010) cleavages.

The lower photograph, taken under crossed polars, shows albite and pericline twin lamellae forming a cross-hatched or 'tartan' pattern which at first sight resembles the texture seen in microcline. In anorthoclase however the lamellae are seen in sections cut nearly perpendicular to the x crystallographic axis as in this case and, in this orientation, a nearly centred acute bisectrix interference figure can be obtained.   [285].

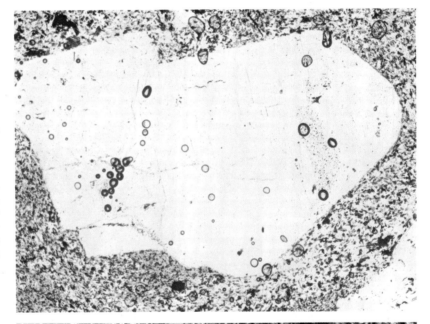

Specimen from pantellerite, Pantelleria, Italy; magnification × 37.

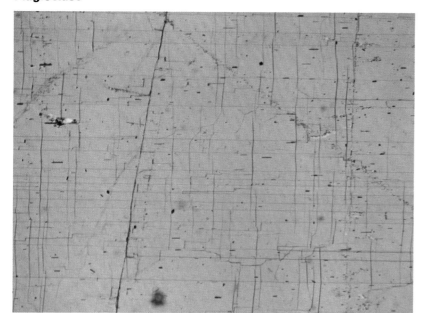

Plagioclase

$NaAlSi_3O_8–CaAl_2Si_2O_8$

Symmetry		=	Triclinic ($+$) or ($-$)
RI	β	=	1·532–1·585
Birefringence		=	0·007–0·013

These photographs are of a thin section of a labradorite crystal cut almost exactly at right-angles to the x crystallographic axis, and all three exposures were made under crossed polars. In the upper photograph the (010) cleavage has been set parallel to the long edge of the photograph which is perpendicular to the vibration direction of the polarizer. The (001) cleavage is not parallel to the short edge of the photograph but shows slight changes in direction at the boundaries of the albite twin lamellae seen in the two lower photographs. The middle and lower photographs show the appearance of the section after rotation to the extinction positions of the two sets of twin lamellae. The angle of rotation in each case is 26°: from a graph of composition in the plagioclase feldspar series against extinction angle in a section cut perpendicular to the x axis, the compositon of this crystal is about $Ab_{50}An_{50}$. [318].

Specimen from unknown locality; magnification $\times 43$.

Plagioclase

$NaAlSi_3O_8$–$CaAl_2Si_2O_8$

Symmetry	=	Triclinic (+) or (−)
RI β	=	1·532–1·585
Birefringence	=	0·007–0·013

These photographs show a number of phenocrysts and microphenocrysts of plagioclase in the fine-grained groundmass of an andesite (a few phenocrysts of orthopyroxene are also visible). In plane-polarized light (upper photograph) zoning can be seen by the arrangement of inclusions in the group of feldspar crystals in the centre of the field. Within this aggregate the brown material is cryptocrystalline groundmass incorporated in the growing crystals.

Under crossed polars (lower photograph) lamellar twinning is visible in most of the crystals and oscillatory zoning is obvious in the crystals which are near to their extinction positions. The low grey interference colours are slightly anomalous due to dispersion, a feature which is not uncommon in plagioclases from volcanic rocks. [318].

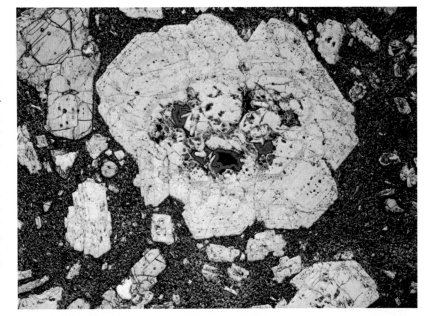

Specimen from pyroxene–andesite, Matra Hills, near Budapest, Hungary; magnification × 26.

Quartz

SiO$_2$

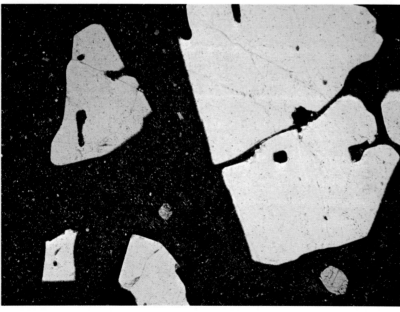

Symmetry		=	Trigonal (+)
RI	ω	=	1·544
	ε	=	1·553
Birefringence		=	0·009

Quartz is the most common of all minerals and is fairly easily distinguished from feldspar in thin section because it is generally unaltered and lacks visible twinning or cleavage. (It may contain fluid inclusions and if these are very small and numerous they may give the quartz a dusty appearance.)

The upper photograph, taken in plane-polarized light, shows clear quartz phenocrysts in a volcanic rock and these have embayments against the groundmass of the rock: this is not an uncommon feature and although sometimes interpreted as due to resorption of the crystals it may be due to the rapidly growing crystal enclosing the groundmass material.

In the lower photograph, taken under crossed polars, some crystals show the white interference colours characteristic of crystals cut nearly parallel to the optic axis. Signs of a yellowish interference colour is an indication that the thin section is slightly too thick. [340].

Specimen from quartz porphyry, Dundubh, Isle of Arran, Scotland; magnification × 21.

Quartz

SiO$_2$

Symmetry	=	Trigonal (+)
RI ω	=	1·544
ε	=	1·553
Birefringence	=	0·009

These photographs show quartz in a metamorphic rock. In the upper photograph, taken in plane-polarized light, most of the field of view appears to be occupied by clear quartz crystals with a few small inclusions. The minerals at the corners of the photograph are biotite and sillimanite.

In the lower photograph, taken under crossed polars, individual crystals of quartz can be seen but, within these, the extinction is not uniform and this shadowy extinction is fairly common in deformed rocks. Closer inspection of this view reveals that, at the top right corner and the bottom left corner of the field of view, there are regions which differ in that the crystals have dark borders and dark veinlets penetrating into the crystal: at the left-hand bottom corner there are signs of twin lamellae in one of the crystals. These are cordierite crystals and they have been included in this field of view to show that the relief and birefringence of cordierite may, depending on its composition, be very similar to that of quartz but can be distinguished by signs of alteration to pinite at the edges of the cordierite grains.   [340].

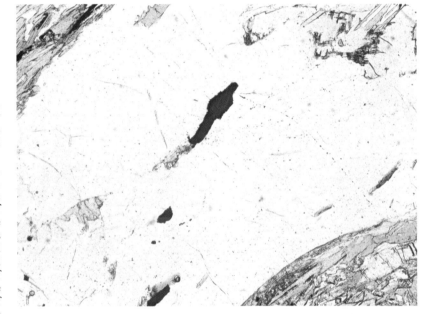

Specimen from cordierite–sillimanite gneiss, 11 km south of Ihosy, Madagascar; magnification × 43.

Myrmekite

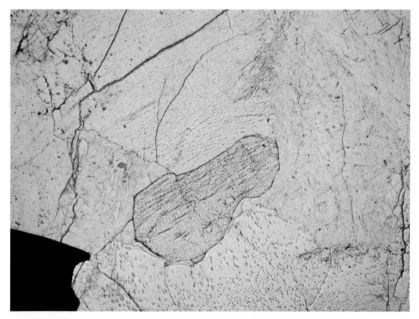

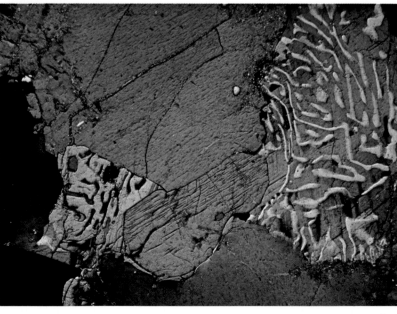

This consists of an intergrowth of plagioclase and quartz with a vermicular texture which is clearly seen in the lower photograph, taken under crossed polars. This should be compared with granophyric texture (q.v.). In the upper photograph, taken under plane-polarized light, the intergrowth is almost invisible because the plagioclase and the intergrown quartz have almost the same RI whereas the remainder of the field, except for an apatite crystal in the centre, is occupied by alkali feldspar with lower RI. The alkali feldspar has a microperthitic texture with oriented inclusions of plagioclase.

Specimen from charnockite, 25 km north-west of Fort Dauphin, Madagascar; magnification × 52.

Granophyric texture

These photographs show an intergrowth of quartz and alkali feldspar. Even in plane-polarized light (upper photograph) the intergrowth is visible because of the difference in refringence of the two minerals and the fact that the alkali feldspar is brown, due to alteration, whereas the areas of quartz are clear. Under crossed polars (lower photograph) what appear to be the outlines of individual crystals are visible. What is not known is what material these shapes represent since each 'crystal' consists of about equal amounts of quartz and feldspar although it is probable that the crystal outline is that of feldspar.

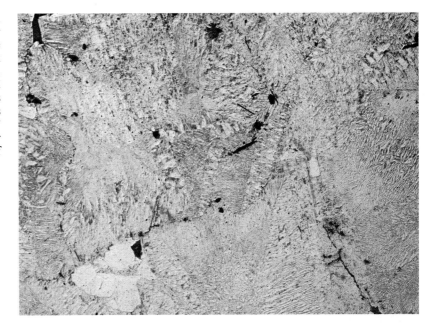

Specimen from granophyre, Eastern Red Hills, Skye, Scotland; magnification × 32.

Tridymite

SiO$_2$

Symmetry	=	Orthorhombic (+)
RI β	=	1·470–1·480
Birefringence	=	0·002–0·004

The upper photograph, taken in plane-polarized light, is of a fine-grained rock in which there is an elongated cavity or vein which is mainly filled with tridymite. To the right of the photograph the tridymite shows up in relief against the mounting medium where there are small holes in the slide.

The lower photograph, taken under crossed polars, shows that the tridymite has very low birefringence: the wedge-shaped twinned crystals are characteristic of this mineral. Although the name tridymite implies three-fold groups of twins, two-twinned individuals are probably more common.   [340].

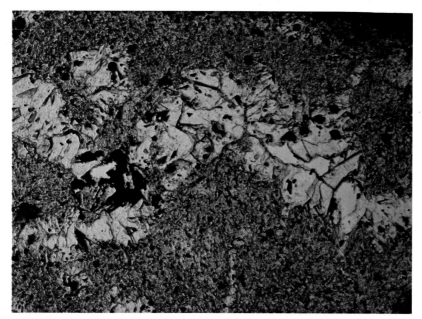

Specimen from dacite, Hakone Volcano, Japan; magnification × 72.

Cristobalite

SiO_2

Symmetry	=	Tetragonal $(-)$	
RI	ε	=	1·484
	ω	=	1·487
Birefringence	=	0·003	

In the upper photograph, taken in plane-polarized light, cristobalite is intergrown with pyroxene (brown), plagioclase feldspar (colourless) and opaque crystals which are probably mainly ilmenite. The cristobalite shows moderate relief against the plagioclase feldspars because of the low index of the cristobalite. It is characterized by what is known as 'tile' structure, i.e. resemblance to curved tiles on a roof. Under crossed polars (lower photograph), the cristobalite shows very low grey colours which are characteristic. The different orientations shown by the differences in birefringence are partly due to the tile structure but multiple twinning is also present.   [340].

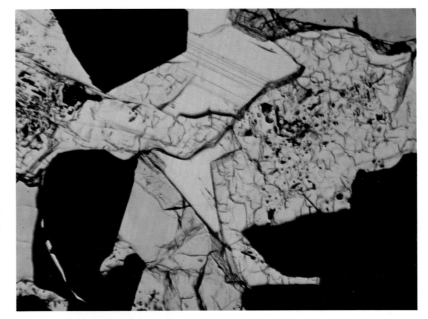

Specimen from coarse-grained basalt, Apollo 17 lunar sample; magnification × 164.

Nepheline

NaAlSiO$_4$

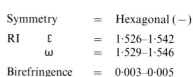

Symmetry		=	Hexagonal (−)
RI	ε	=	1·526–1·542
	ω	=	1·529–1·546
Birefringence		=	0·003–0·005

The colourless minerals in the upper photograph, taken in plane-polarized light, are mainly nepheline and cancrinite. A slight difference in relief can just be detected but it is necessary to look at the lower photograph (taken under crossed polars) to see clearly that the areas with bright interference colours consist of cancrinite (q.v.) while the very dark areas are of nepheline. No cleavages are visible so that there is unlikely to be any alkali feldspar in this field of view. Cancrinite is very often associated with nepheline particularly in plutonic rocks and this gives a clue to the presence of nepheline.

The green mineral in these photographs is aegirine-augite and the occurrence of an alkali pyroxene indicates that the rock is rich in alkalis but is not necessarily nepheline-bearing. [356].

Sample from nepheline syenite, Khabozero, Kola, USSR; magnification × 34.

Nepheline

$NaAlSiO_4$

Symmetry		=	Hexagonal $(-)$
RI	ε	=	1·526–1·542
	ω	=	1·529–1·546
Birefringence		=	0·003–0·005

These photographs show nepheline phenocrysts in the fine-grained groundmass containing also small green crystals of aegirine-augite and a few crystals of sphene. It is common to find nepheline phenocrysts together with alkali feldspar phenocrysts in the same rock and it is sometimes difficult to distinguish them (see photographs on p.78). All colourless phenocrysts in this field of view are of nepheline although there are some small sanidines in the groundmass. The crystal at the top left of the field of view with the inclusion of aegirine-augite shows a hexagonal outline but is incomplete.

In the lower photograph, taken under crossed polars, the hexagonal-shaped crystal is fairly dark and is cut nearly perpendicular to the optic axis. A few small sanidine crystals can be recognized in the groundmass by the presence of simple twinning.   [356].

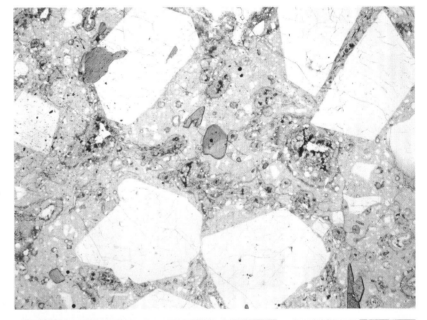

Specimen from phonolite, ejected block, Oldoinyo Lengai, Tanzania; magnification × 12.

Sanidine & Nepheline

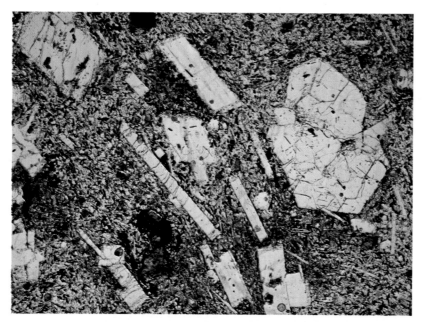

The upper photograph, taken in plane-polarized light, shows phenocrysts of nepheline and sanidine in a groundmass composed mainly of the same two minerals with some pyroxene and a few minute crystals of nosean. Because their relief and birefringence are very similar it is difficult to distinguish nepheline from sanidine. Crystals which show a good cleavage or simple twinning are almost certainly sanidine: basal sections of nepheline are hexagonal and prismatic sections are usually nearly square so that the two narrow elongated crystals in the centre of the field are sanidines, whereas the group of crystals to the right of the field showing partly hexagonal shape are nepheline.

Under crossed polars (lower photograph) a few crystals show simple twins: these are sanidines twinned on the Baveno law. In the group of crystals to the right of the field, one of the crystals showing hexagonal shape is almost black and is nepheline cut almost exactly perpendicular to the optic axis, the other is cut slightly oblique to the optic axis and shows very low birefringence with slight zoning near the margins.   [285] [356].

Specimen from nosean phonolite, Wolf Rock, Cornwall, England; magnification × 21.

Leucite

$KAlSi_2O_6$

Symetry	=	Tetragonal (Pseudocubic) (+)
RI n	=	1·508–1·511

The upper photograph, taken in plane-polarized light, shows colourless phenocrysts of leucite together with microphenocrysts of aegirine-augite (greenish-brown) and nosean (zoned brown crystals) in a fine-grained groundmass mainly of plagioclase but containing small leucite and nosean crystals. Leucite sometimes has radially or concentrically arranged inclusions of glass but this is not apparent here.

Under crossed polars (lower photograph) the appearance of multiple twinning in more than one orientation is very characteristic of leucite: very small leucite crystals in the groundmass appear isotropic. Leucite is sometimes replaced by an intergrowth of alkali feldspar and nepheline while retaining the shape of the leucite crystals and this is termed pseudoleucite.   [367].

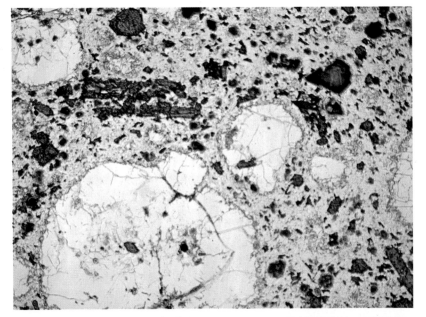

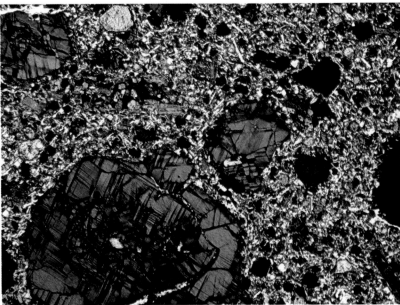

Specimen from leucitophyre, Reiden, Eifel, Germany; magnification × 20.

Nosean

$6NaAlSiO_4 \cdot Na_2SO_4$

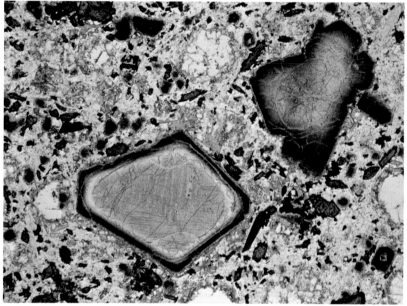

Symmetry		=	Cubic
RI	n	=	1·495

These photographs are from different rocks and both are taken in plane-polarized light only.

The upper photograph shows two nosean phenocrysts with dark, almost black, rims due to iron oxide inclusions: throughout the cores of the crystals there are numerous orientated inclusions. A few microphenocrysts of leucite are visible in the field and this is the same rock as used to illustrate leucite (q.v.); there are also microphenocrysts of nosean, dark brown with inclusions similar to the rims of the phenocrysts, green to brown pyroxene in a fine groundmass mainly of plagioclase with some calcite: due to stray polarization in the microscope the calcite shows weak interference colours in this photograph.

The lower photograph shows nosean crystals, full of inclusions, intergrown with sanidine which is free from inclusions. The sub-stage aperture has been opened wide but closing it allows one to see the difference in relief between the nosean and the sanidine – nosean has a much lower RI and is isotropic. [375].

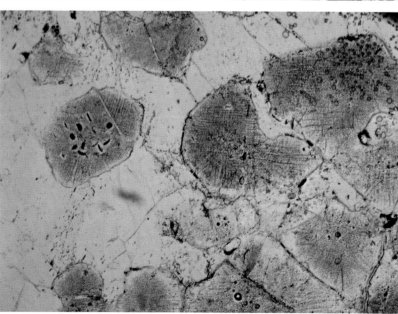

Upper specimen from leucitophyre, Reiden, Eifel, Germany; magnification × 27. Lower specimen from nosean sanidinite, Laacher See, Germany; magnification × 53.

Cancrinite

$6NaAlSiO_4 \cdot Na_2CO_3$

Symmetry	=	Hexagonal $(-)$
RI ε	=	1·503
ω	=	1·528
Birefringence	=	0·025

In the upper photograph, taken in plane-polarized light, the objective was raised slightly to show the Becke line. The minerals which have the Becke line within their boundaries are nepheline and alkali feldspar which in this rock have almost identical refractive indices – the low refractive index mineral is cancrinite.

Under crossed polars (lower photograph) the cancrinite shows first-order colours, except for the large crystal to the left of the field which is a low second-order bluish-red colour. In nepheline syenites the appearance of a colourless mineral with first- or second-order interference colours and a very low refractive index is usually a good indication of the presence of cancrinite and in turn this leads the observer to look for the presence of nepheline. [381].

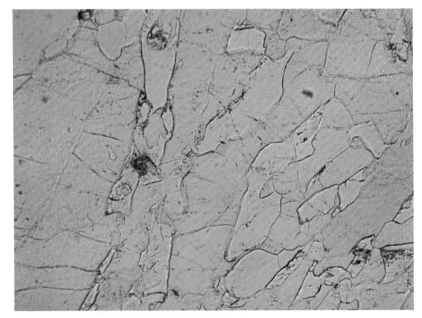

Specimen from nepheline–syenite, unknown locality; magnification × 72.

Scapolite

$$3(NaAlSi_3O_8) \cdot NaCl - 3(CaAl_2Si_2O_8) \cdot CaCo_3$$

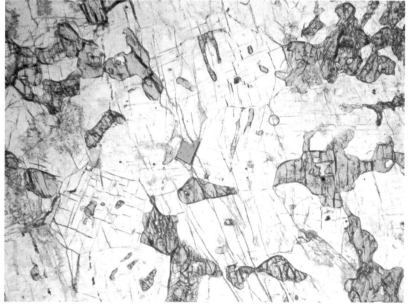

Symmetry		=	Tetragonal $(-)$
RI	ε	=	1·540–1·564
	ω	=	1·546–1·600
Birefringence		=	0·005–0·038

The upper photograph, taken in plane-polarized light, shows mainly scapolite (colourless) together with a pale green clinopyroxene and one crystal of biotite in the centre of the field. A crystal at the lower left part of the field shows two sets of cleavages at right-angles to one another. Most of the other crystals show at least one cleavage.

In the lower photograph, taken under crossed polars, some of the crystals show second-order colours but those showing two cleavages have low interference colours and are cut nearly at right-angles to the optic axis since the $\{100\}$ cleavages are parallel to the optic axis. The composition in the scapolite series, from the marialite end-member (Na-rich) to the meionite end-member, is obtained from either a refractive index determination or a measurement of birefringence.   [384].

Specimen from edge of phlogopite–pegmatite, near Mafilefy, Madagascar; magnification × 20.

Analcite

$NaAlSi_2O_6 \cdot H_2O$

Symmetry		=	Cubic
RI	n	=	1·479–1·493

The upper photograph, taken in plane-polarized light, shows a triangular area of analcite bounded by elongated crystals of plagioclase. The refractive index of analcite is very low so that, with the sub-stage diaphragm closed, it stands out in relief against the surrounding minerals.

The lower photograph, taken under crossed polars, shows that the analcite is isotropic. Sometimes analcite shows dark grey interference colours and complex twinning and in such cases it can be mistaken for leucite. In some rocks leucite may be entirely replaced by analcite but rarely are the two minerals seen together. It may be necessary to measure the refractive index to distinguish analcite from leucite.   [389].

Specimen from crinanite, Howford Bridge, Ayrshire, Scotland; magnification × 62.

Corundum

Al_2O_3

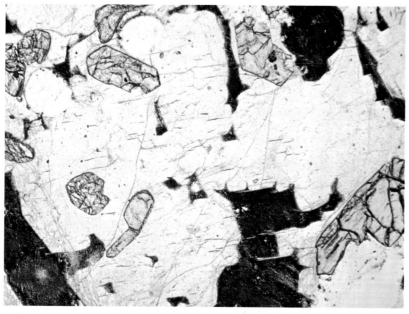

Symmetry		=	Trigonal $(-)$
RI	ε	=	1·760–1·763
	ω	=	1·768–1·772
Birefringence		=	0·008–0·009

The upper photograph shows a number of corundum crystals embedded in feldspar. Its high relief is characteristic and when it shows a slight bluish colour, as it does here, this is an indication of the presence of sapphire (blue variety of corundum). The almost black material is glass full of inclusions.

In the lower photograph, taken under crossed polars, the highest colours seen are first-order yellow. Because of its extreme hardness corundum crystals may be thicker than the surrounding minerals and so show slightly higher colours than the birefringence indicates. Multiple twinning is quite common in corundum but is not present in any of these crystals. [405].

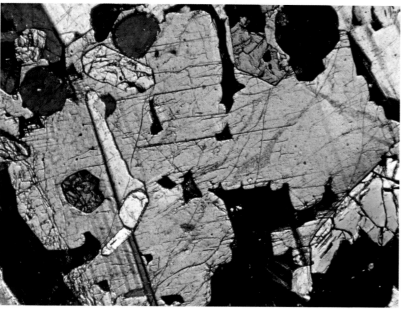

Specimen from buchite, Rudh' a' Chromain Sill, Ross of Mull, Scotland; magnification × 52.

Rutile

TiO_2

Symmetry		=	Tetragonal (+)
RI	ω	=	2·605–2·613
	ε	=	2·899–2·901
Birefringence		=	0·286–0·296

The upper and middle photographs show some fairly large crystals of rutile in a mass of altered plagioclase feldspar. One of the crystals shows two good cleavages at an angle of approximately 60°. The deep golden-brown colour is fairly characteristic of rutile. Under crossed polars (middle photograph) it can be seen that there are twin lamellae parallel to the traces of the two cleavages, which are probably {011}. This crystal has been set near to the extinction position to show the twinning. Because of the strong absorption colour it is not possible to estimate the birefringence which is very high nor to be aware of the very high refractive indices. The other crystal which does not show twinning appears to be the same colour in plane light and under crossed polars.

The lower photograph, taken in plane-polarized light, shows needles of rutile within biotite. The occurrence of rutile as needles in biotite and quartz is fairly common but when the needles are as fine as those illustrated here there is very little that can be done optically to establish that they are indeed rutile. [415].

Upper and middle specimen from altered anorthosite, Roseland, Virginia, USA; magnification × 20. Lower specimen from unknown locality; magnification × 72.

Perovskite

CaTiO$_3$

Symmetry = Monoclinic (pseudo-cubic) (+)

RI n = 2·30–2·38

The upper photograph, taken in plane-polarized light, shows a few dark brown crystals of perovskite intergrown with melilite (colourless) and iron ore (black). Zoning of the brown colour of the perovskite can be seen. The relief is very high but the strong absorption colour tends to obscure this.

In the lower photograph, taken under crossed polars, the perovskite crystals are birefringent and show complex multiple twinning. This cross-hatched twinning is a characteristic of perovskite and serves to distinguish it from some other dark brown minerals. Its occurrence along with melilite is common (see description of melilite, p.29). [422].

Specimen from melilite rock, Scawt Hill, County Antrim, Ireland; magnification × 43.

Spinel

(Fe, Mg) Al$_2$O$_4$

Symmetry	=	Cubic
RI n	=	1·719–1·835

The spinel group covers a wide range of chemical composition but the common varieties are aluminous with Fe and Mg substitution. The range of RI given does not cover the ferric iron and chromium-rich varieties. The characteristic colours are dark green or dark brown and the two microphotographs, both taken in plane-polarized light, show two different occurrences.

The upper photograph shows dark green spinel together with olivine. The spinel crystals are subhedral in shape and their colour is zoned so that some of the crystals show brownish cores.

In the lower photograph the spinel is an even darker olive-green colour. The shapes of the crystals are determined by the calcic plagioclase with which it is intergrown. Some of the dark regions in this field of view consist of glass crowded with inclusions but these can be readily distinguished from the spinel because of the high relief of the spinel. [424].

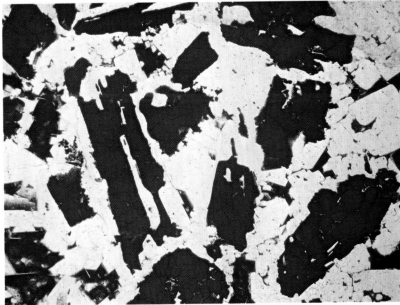

Upper specimen from spinel–forsterite–xenolith, Vesuvius, Italy; magnification × 43. Lower specimen from buchite, Rudh' a' Chromain Sill, Ross of Mull, Scotland; magnification × 20.

Brucite

$Mg(OH)_2$

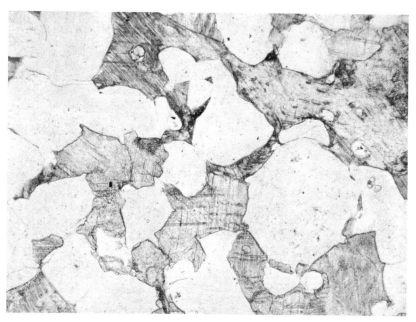

Symmetry		=	Trigonal (+)
RI	ω	=	1·560–1·590
	ε	=	1·580–1·600
Birefringence		=	0·012–0·020

The upper photograph, taken in plane-polarized light, shows brucite (clear areas) intergrown with dolomite (darker areas). These are probably pseudomorphs after periclase (MgO).

In the view under crossed polars (lower photograph), the brucite areas consist of aggregates of fibres with low birefringence. A few regions which show anomalous blue colours are of serpentine. [434].

Specimen from brucite marble, Ledbeg, Assynt, Scotland; magnification × 72.

Calcite

CaCO$_3$

Symmetry		=	Trigonal ($-$)
RI	ε	=	1·486
	ω	=	1·658
Birefringence		=	0·172

Most of the field of view is occupied by calcite and the upper and middle photographs taken in plane-polarized light show the change in relief produced by rotating the polarizer through 90°. This is referred to as 'twinkling' and is most easily seen by rotating the polarizer as has been done here. Because of the perfect rhombohedral cleavage most crystals show at least one good cleavage.

Under crossed polars (lower photograph) the interference colours can be seen to be of very high order: this section may be slightly less than the normal 0·03 mm in thickness since in sections of standard thickness the interference colour produced is a high-order white. Twinning can be seen in a few crystals and this can be useful in distinguishing calcite from dolomite (q.v.). [476].

Specimen from diopside–forsterite marble, Loch Duich, Scotland; magnification × 44.

Dolomite

$CaMg(CO_3)_2$

Symmetry		=	Trigonal $(-)$
RI	ε	=	1·500
	ω	=	1·679
Birefringence		=	0·179

This section contains both calcite and dolomite and, since the two minerals are difficult to distinguish the thin section has been stained. Most of the staining techniques used depend on the fact that calcite is readily soluble in dilute HCl whereas dolomite is not, so that the material which is stained red in these photographs is the calcite. In the upper and middle photographs, taken in plane-polarized light, it can be seen that there are lamellae of dolomite within the stained calcite. The change in relief of the dolomite caused by rotating the polarizer through 90° is clearly shown.

The lower photograph (taken under crossed polars) shows the high birefringence associated with dolomite. The crystal showing a yellow colour to the right of the field of view is forsterite. [489].

Specimen from forsterite marble, Sri Lanka; magnification × 43.

Dolomite

CaMg(CO$_3$)$_2$

Symmetry	=	Trigonal ($-$)
RI ε	=	1·500
ω	=	1·679
Birefringence	=	0·179

These photographs were taken from an unstained section of the same rock used for the previous photographs of dolomite. In the upper photograph (taken under plane-polarized light) two rhombohedral cleavages can be seen at 120° in the crystal in the upper part of the field.

In the lower photograph, taken under crossed polars, twin lamellae are seen to bisect the obtuse angle between the cleavages, i.e. parallel to the short diagonal of the rhomb shape formed by the cleavages. Both dolomite and calcite may have twin lamellae parallel to the long diagonal and parallel to the rhombohedral cleavages themselves, but only dolomite has twin lamellae in the position shown here. [489].

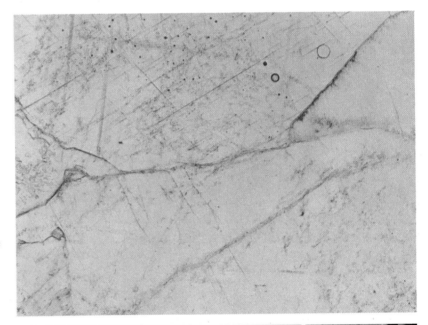

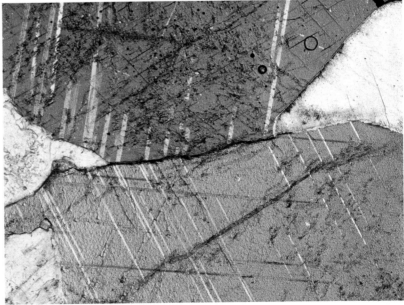

Specimen from forsterite marble, Sri Lanka; magnification × 22.

Apatite

$Ca_5(PO_4)_3(OH, F, Cl)$

Symmetry	=	Hexagonal $(-)$
RI ε	=	1·624–1·666
ω	=	1·629–1·667
Birefringence	=	0·001–0·007

In the upper photograph, taken in plane-polarized light, needles and small hexagonal crystals of apatite stand out in relief against nepheline (clear): the opaque mineral is ilmenite, some of the ilmenite crystals having rims of sphene.

The lower photograph, taken under crossed polars, shows that the birefringence of apatite is about the same as that of the nepheline, and this taken along with the high RI is useful for identification. The one grain showing a blue-green interference colour is aegirine-augite.   [504].

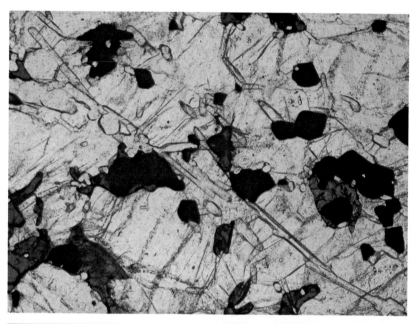

Specimen from apatite–nepheline rock, Lovozero, Kola, USSR; magnification × 40.

Fluorite

CaF$_2$

Symmetry	=	Cubic
RI n	=	1·433–1·435

These two photographs were both taken in plane-polarized light. The upper photograph shows a number of purple crystals of fluorite intergrown with a rare mineral weberite. A few crystals at the top of the photograph show signs of the perfect {111} cleavage. Fluorite has the lowest refractive index of all the common minerals and hence shows considerable relief against most other minerals. In this case the purple colour gives a useful clue to its identity.

The lower photograph shows fluorite as small anhedral grains intergrown with muscovite. Here again the pale purple colour which is unevenly distributed is useful for identification: its isotropic character and very low refractive index would confirm the identification. [511].

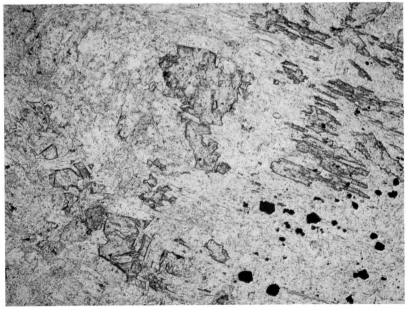

Upper specimen from cryolite deposit, Ivigtut, West Greenland; magnification × 32. Lower specimen from granite, Rostowrack, Cornwall, England; magnification × 44.

Deerite

$$Fe_{12}^{+2}Fe_6^{+3}Si_{12}O_{40}(OH)_{10}$$

Symmetry	=	Monoclinic
RI β	=	1·85
Birefringence	=	0·03 (approx.)

The upper photograph, taken in plane-polarized light, shows needle-shaped deerite crystals intergrown with quartz. The brown mineral at the corners of the field is stilpnomelane. the deerite crystals are almost black but the thin edges of some of the crystals are slightly transparent and show a brown colour. There is a suggestion of a diamond shape in some of the sections due to the development of the $\{110\}$ form.

Under crossed polars (lower photograph) no interference colours can be seen because of the intense absorption.

Specimen from metamorphosed siliceous ironstone, Laytonville, California, USA; magnification × 43.

Howieite

Na(Fe, Mn)$_{10}$(Fe, Al)$_2$Si$_{12}$(O, OH)$_{44}$

Symmetry		=	Triclinic (−)
RI	β	=	1·720
Birefringence		=	0·033

Most of the field of view is occupied by howieite crystals and the pleochroism from yellow to green to a lilac-grey can be seen in different crystals by comparing the upper and middle photographs. In most of the orientations present, there are signs of cleavage.

Under crossed polars (lower photograph), the interference colours are second-order, but are masked to some extent by the absorption colours. The almost black mineral intergrown with howieite is deerite.

Specimen from metamorphosed siliceous ironstone, Laytonville, California, USA; magnification × 24.

Zussmanite

K(Fe, Mg, Mn)$_{13}$Al$_2$Si$_{17}$(O, OH)$_{56}$

Symmetry		=	Trigonal (−)
RI	ε	=	1·623
	ω	=	1·643
Birefringence		=	0·020

Comparison of the upper and middle photographs, taken in plane-polarized light, shows zussmanite, which is pleochroic from pale yellow to pale green, occupying most of the field. A very good cleavage is present in some crystals and the relief against quartz is high.

The lower photograph, taken under crossed polars, reveals that the crystals showing the lowest interference colours are those not showing a cleavage. Sections cut nearly parallel to the perfect cleavage show uniaxial interference figures.

The brown mineral accompanying zussmanite is stilpnomelane.

Specimen from metamorphosed siliceous ironstone, Laytonville, California, USA; magnification × 43.

Yoderite

$Al_3MgSi_2O_8(OH)$

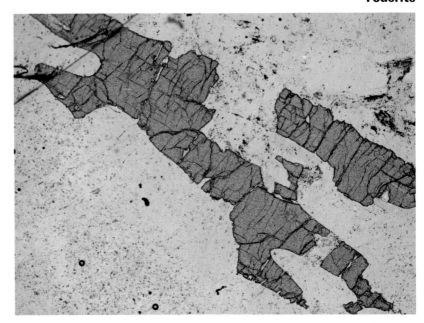

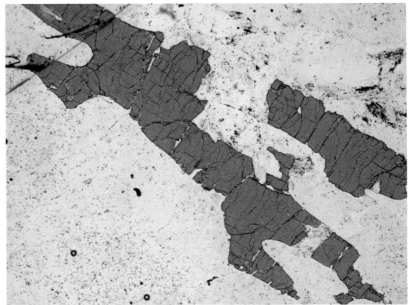

Symmetry	=	Monoclinic (+)
RI β	=	1·691
Birefringence	=	0·026

The spectacular pleochroism of yoderite from purple to a brownish colour can be seen by comparing the two photographs taken in plane-polarized light (upper and middle photographs). Its relief against the surrounding quartz is high. The purple colour is similar to that seen in glaucophane or crossite.

In the lower photograph, taken under crossed polars, the interference colour seen is a combination of the absorption colour and a colour close to first-order red. Because only one orientation is represented here we cannot see the full range of pleochroism or interference colours.

In this rock the yoderite occurs along with quartz and talc and yoderite crystals usually contain a core of kyanite which is not seen in this section.

Specimen from quartz–kyanite–talc schist, Mautia Hill, Tanzania; magnification × 32.

Index

Actinolite   45
Aegirine-augite   39
Aenigmatite   51
Allanite   25
Analcite   83
Andalusite   13-15
Anorthoclase   67
Anthophyllite   43
Apatite   92
Arfvedsonite   50
Astrophyllite   52
Augite   36-7
Axinite   34

Biotite   55-6
Brucite   88

Calcite   89
Cancrinite   81
Chlorite   60-1
Chloritoid   19
Chondrodite   4-5
Cordierite   30-1
Corundum   84
Cristobalite   75
Cummingtonite   44

Deerite   94
Dolomite   90-1

Epidote   23
Eudialyte   21

Fayalite   1, 2
Ferroactinolite   45
Fluorite   93

Forsterite   1, 2

Garnet   8
Gedrite   43
Glaucophane   49
Granophyric texture   73
Grunerite   44

Hornblende   46-7
Howieite   95

Idocrase   9

Jadeite   40

Kaersutite   48
Kyanite   16

Lamprophyllite   53
Lawsonite   26
Leucite   79

Melilite   28-9
Microcline   64
Monticellite   3
Mullite   12
Muscovite   54
Myrmekite   72

Nepheline   76-8
Nosean   80

Olivine   1-2
Orthite   25
Orthopyroxene   35

Pectolite   42

Perovskite   86
Perthite and microperthite
Piemontite   24
Plagioclase   68-9
Prehnite   63
Pumpellyite   27
Pyrophyllite   58

Quartz   70-1

Rutile   85

Sanidine   66, 78
Sapphirine   20
Scapolite   82
Serpentine   62
Sillimanite   10-11, 15
Sphene   7
Spinel   87
Staurolite   18
Stilpnomelane   57

Talc   59
Topaz   17
Tourmaline   32-3
Tremolite   45
Tridymite   74

Vesuvianite   9

Wollastonite   41

Yoderite   97

Zircon   6
Zoisite   22
Zussmanite   96